Building Quality
Management Systems

Selecting the Right Methods and Tools

Building Quality
Management Systems

Selecting the Right Methods and Tools

Luis Rocha-Lona
Jose Arturo Garza-Reyes
Vikas Kumar

CRC Press
Taylor & Francis Group
Boca Raton   London   New York

CRC Press is an imprint of the
Taylor & Francis Group, an **informa** business

CRC Press
Taylor & Francis Group
6000 Broken Sound Parkway NW, Suite 300
Boca Raton, FL 33487-2742

© 2013 by Taylor & Francis Group, LLC
CRC Press is an imprint of Taylor & Francis Group, an Informa business

No claim to original U.S. Government works

Printed on acid-free paper
Version Date: 20130424

International Standard Book Number-13: 978-1-4665-6499-2 (Paperback)

Library of Congress Cataloging-in-Publication Data

Rocha-Lona, Luis.
    Building quality management systems : selecting the right methods and tools / Luis Rocha-Lona, Jose Arturo Garza-Reyes, Vikas Kumar.
    pages cm
    Includes bibliographical references and index.
    ISBN 978-1-4665-6499-2 (pbk. : alk. paper)
    1. Total quality management. 2. Strategic planning. 3. Industrial management. I. Title.

HD62.15.R63293 2013
658.4'013--dc23   2013014908

Visit the Taylor & Francis Web site at
http://www.taylorandfrancis.com

and the CRC Press Web site at
http://www.crcpress.com

We deeply thank our families for supporting us in our efforts to complete this book, as that meant putting up with an overworked family member during nights, weekends, and holidays. This work is dedicated to them.

Contents

Preface

Quality has become one of the most important decision-making factors for customers purchasing a specific product or selecting a service. Many organizations invest a considerable amount of human resources, capital, and time to build the right quality management systems (QMSs). In many instances, however, QMSs and the adoption of specific business and quality improvement models, methods, and tools are not adequate or are poorly deployed. In addition, in many cases, QMSs are not aligned with strategic quality planning and business strategies. This misalignment results in inadequate implementations that produce severe pitfalls and frustration for these organizations. Thus, the purpose of this book is to help directors, practitioners, consultants, researchers, and all kinds of professionals make effective decisions in relation to the design, implementation, and improvement of QMSs. In addition, the book aims at helping all professionals set a strategic quality plan in terms of their organizations' specific needs, capabilities, cost–benefits, policies, and business strategies.

The book is based on our industrial experience as consultants, researchers, and academics, after working on several business improvement projects for multinational organizations that wanted to design, implement, or improve their QMSs. Our experience made us realize that most QMS implementation problems are the result of a lack of specific methodologies that clearly indicate to the organizations the steps that they need to follow to successfully deploy their QMSs. This encouraged us to work on this book, *Building Quality Management Systems: Selecting the Right Methods and Tools*, to assist professionals in making better decisions while developing and deploying QMSs in their organizations.

The first two chapters of the book provide an overview of QMSs and systems thinking, the relevance of QMSs and their impact on competitiveness and financial performance, and the most well-known business and quality improvement models, methods, and tools. Chapter 3 reviews the process management approach, which we consider to be an essential element in supporting an organization's QMS. Altogether, Chapters 4 through 7 present the methodology that we propose for an organization to design, implement, or enhance its QMS. The proposed methodology consists of evaluating the organization's QMS and business processes (Chapter 4); strategically planning and aligning its improvement agenda with the business strategy (Chapter 5); selecting the right models, methods, and tools to be adopted as part of its QMS (Chapter 6); understanding the QMS implementation challenges and critical success factors (Chapter 7); and evaluating such implementation (Chapter 8). Each of these chapters is intended to clearly indicate to the reader how to carry out all the activities that comprise the stages of our methodology. Finally, Chapter 9 highlights the importance of quality as a way of life and the opportunity that it presents for organizations to enhance their competitiveness.

Acknowledgments

We would like to thank the National Polytechnic Institute of Mexico (IPN) for sponsoring this project, and the Centre for Supply Chain Improvement, Derby Business School, the University of Derby and the Dublin City University Business School, Dublin City University for the support received to produce this work. We would also like to thank our publisher, Productivity Press, and our executive editor, Michael Sinocchi, for all the support in completing this book.

About the Authors

Dr. Luis Rocha-Lona has over 10 years of work experience in the public and private sectors. He holds a PhD in operations management from Manchester Business School at the University of Manchester in the UK, and an MSc in control systems with a major in information systems/manufacturing at the University of Sheffield, UK. Dr. Rocha-Lona graduated from the National Polytechnic Institute of Mexico as an automation control systems engineer. He has led several research projects sponsored by private companies and the Mexican government through the National Council of Science and Technology (CONACYT) and is actively involved in consulting activities to manufacturing and service organizations.

Dr. Rocha-Lona joined the Business School at the National Polytechnic Institute of Mexico in 2007, where he is a senior lecturer in operations management and quality management systems. He has presented his work in several international venues and congresses and serves as a member of the scientific committees for multiple international conferences. He is a member of the Institute of Operations Management (IOM) and the American Society for Quality (ASQ). His current research interests are in the areas of performance measurement systems, business process improvement, and business strategy.

Dr. Jose Arturo Garza-Reyes is a senior lecturer in operations and supply chain management at the Centre for Supply Chain Improvement, Derby Business School, the University of Derby, UK. He holds a PhD in manufacturing systems and operations management from Manchester Business School at the University of Manchester (UK), an MBA from the University of Northampton (UK), an MSc in production and quality from the Autonoma de Nuevo Leon University (Mexico), a postgraduate certificate in teaching and learning in higher education from the University of Derby (UK), and a BSc in mechanical management engineering from the Autonoma de Nuevo Leon University (Mexico).

He has published a number of articles in leading international journals and conferences as well as a book about manufacturing performance measurement systems. Dr. Garza-Reyes has participated as a guest editor for special issues in the *International Journal of Lean Enterprise Research* (*IJLER*), the *International Journal of Engineering Management and Economics* (*IJEME*), and the *International Journal of Engineering and Technology Innovation* (*IJETI*). He currently serves on the editorial board of several international journals and has contributed as a member of the scientific and organizing committees of several international conferences.

His research interests include general aspects of operations and manufacturing management, operations and quality improvement, and supply chain improvement. Dr. Garza-Reyes is a Chartered Engineer (CEng), a certified Six Sigma Green Belt, and has over six years of industrial experience working as production manager, production engineer, and operations manager for several international and local companies in both the UK and Mexico. He is also a member of the Institution of Engineering Technology (IET) and a fellow member of the Higher Education Academy (FHEA).

Dr. Vikas Kumar has over six years of experience in area operations management. He holds a PhD in management studies from Exeter Business School, UK, and a bachelor of technology (first-class distinction) in metallurgy and materials engineering from the National Institute of Foundry and Forge Technology (NIFFT, Ranchi) in India. He also holds the status of associate of the Higher Education Academy (AHEA). Dr. Kumar joined Dublin City University Business School as a lecturer in management in 2009. He previously worked as a research assistant at the University of Hong Kong, and was a visiting scholar at the Indian Institute of Management, Ranchi, in India and at Khon Kaen University, Nong Khai campus, in Thailand. He has worked on a number of consultancy projects for many multinational firms, such as BT, EDF Energy, LTSB, and Vodafone. He is also actively involved in process improvement projects in Irish hospitals.

Dr. Kumar has contributed to many book chapters and has been published in leading journals such as the *International Journal of Production Research*, *Expert Systems with Applications*, *Strategic Change*, and *Computers and Industrial Engineering*. He serves on the editorial board of four international journals and has participated as a guest editor for special issues of *Production Planning and Control*, *International Journal of Lean Enterprise Research (IJLER)*, *International Journal of Engineering Management and Economics (IJEME)*, and *International Journal of Engineering and Technology Innovation (IJETI)*. His current research interests include process modeling, healthcare management, supply chain management, service operations management, and operations strategy.

Abbreviations

ABEF: Australian Business Excellence Framework
AS: Aerospace Standards (AS-9100 for the aerospace industry)
BEM: Business excellence model
BIS: Business information system
BP: Business process
BPIR: Business performance improvement resource (model)
BPM: Business process management
BPR: Business process reengineering
BSC: Balanced scorecard
BS: British Standards
CAD: Computer-aided design
CAM: Computer-aided manufacturing
CapUT: Capacity utilization
CFBE: Canadian Framework for Business Excellence
CI: Continuous improvement
CRM: Customer relationship management
CSF: Critical success factor
CSM: Current state mapping
DFSS: Design for Six Sigma
DMAIC: Define, measure, analyze, improve, control
DOE: Design of experiments
DSS: Decision support system
EFQM: European Foundation for Quality Management
EQA: European Quality Award
ERP: Enterprise resource planning
FMEA: Failure mode and effects analysis
FSM: Future state mapping
IATF: International Automotive Task Force
IDEF0:   Icam DEFinition Zero

IMSS: Integrated Management System Standards
IP: Internet Protocol
ISO: International Organization for Standardization
IT: Information technology
JIT: Just-in-time
KM: Knowledge management
KPI: Key performance indicator
MDI: Maturity diagnostic instrument
MIS: Management information system
MQMC: Mexican quality model for competitiveness
MRP: Material requirements planning
NIST: National Institute for Science and Technology
OEE: Overall equipment effectiveness
OHSAS: Occupational Health and Safety Management System
(for the certification of a health and safety system)
OP: Overall productivity
PAF: Prevention, appraisal, failure
PCA: Process capability analysis
PDCA: Plan, do, check, act
POLCA: Paired-cell overlapping loops of cards with authorization
POS: Operational pharmaceutical site
QC: Quality circle
QFD: Quality function deployment
QI: Quality improvement
QM: Quality management
QMI: Quality management initiative
QMS: Quality management system
QRM: Quick response manufacturing
QS: Quality Standards (QS-9000 for the automotive industry)
SAE: Society of Automotive Engineers
SIPOC: Supplier, inputs, process, outputs, customers
SMEs: Small- and medium-sized enterprises
SMED: Single-minute exchange of die
SPC: Statistical process control
SQA: Singapore Quality Award (framework)
SQP: Strategic quality planning
TCP: Transmission Control Protocol
TOC: Theory of constraints (international)
TPM: Total productive maintenance

TPS: Toyota Production System
TPT: Throughput time
TQ: Total quality
TQM: Total quality management
TS: TS-29001 (for technical specification for the petroleum, petrochemical, and natural gas industries)
VE: Virtual enterprise
VOC: Voice of the customer
VSM: Value stream mapping

Chapter 1

Introduction

1.1 Knowledge-Based Economies, Competitiveness, and Innovation

We live in a world driven by the large-scale production of goods, which becomes more demanding as world population grows and becomes sophisticated. The real issue for most businesses is that they face competitive markets that change rapidly due to economic, political, sociocultural, and technological factors. The challenge today for any business is to maximize its profits while also maximizing customer value in a sustainable way. This of course is not an easy job and requires a business management system that considers the full organization of every process to deliver high-quality products and services to its customers. The challenge today is how businesses are managed to maximize customer value in a rapidly changing environment (Cobb, 2001).

Knowledge-based economies are characterized as being sophisticated in the way they produce, deliver, and consume products and services. They have evolved from industrialist economies, and they have specific needs to be covered considering sustainability, government legislations, technology, and social responsibility. In this context, the challenge is to produce goods of high quality. This requires from organizations considerable amounts of investments dedicated to building and developing quality management systems (QMSs) that address those demands. In addition, it is necessary that organizations adapt quickly to unpredicted changes and markets trends. For industries such as high-tech, pharmaceutical, financial, automotive, and energy, these issues are even more critical. Product cycles for those organizations have reduced dramatically, allowing no room for mistakes when planning, designing, producing, and delivering their goods. Therefore, understanding this dynamic environment in knowledge-based economies is essential for organizations aiming to play a key role in their industries.

1.1.1 Quality Management Systems

We can define a quality management system (QMS) as *an integrated business approach to plan and deploy quality management models, methods, and tools across the organization with alignment to business strategy.* The elements that compose a QMS can be categorized into human capital, processes, management models, methods and tools, business strategy, and information technology. Many companies are aiming to become world-class organizations and to achieve business excellence through the strategic implementation of QMSs. This sounds like something organizations and business people would like to achieve. The good news is that we have accumulated business knowledge that can help to achieve this goal. However, business knowledge accumulated since the industrial revolution is only an element that can help—we certainly require more than that. It is necessary that organizations consider their customer needs and their own resources in order to strategically plan, develop, deploy, and evaluate their QMSs.

1.1.2 QMSs and Competitive Advantage

Many organizations deploy considerable efforts to become competitive in their industries. Some QMSs have their origin in the competitiveness area (e.g., the Malcolm Baldrige model and the EFQM model). The U.S. launched the Malcolm Baldrige National Quality Program in 1987,[*] commonly known as the Baldrige model. Soon afterwards, in 1991, the European Foundation for Quality Management introduced the EFQM model. These models appeared with the objective of guiding organizations' improvements in their business excellence journey, and were originated so that organizations were competitive. Governments' efforts at these times were focused on motivating organizations to rethink and redesign business models, quality methods, and tools to recover lost markets and develop new ones. As a result, QMSs

---

[*] The program was launched under the Malcolm Baldrige National Quality Improvement Act of 1987 to encourage U.S. firms' competitiveness. See Commerce, U.S.O. (1987). Malcolm Baldrige National Quality Improvement Act.

were developed and linked to competitiveness and business models. These business excellence models (BEMs) are reviewed later on in this book, as they provide the elements and business criteria to manage organizations as a whole business.

Porter (1998) defines competitiveness as the ability of organizations to stay in the market, playing a key role in their industry. However, competitiveness should not be the last objective of quality management systems. QMSs are able to contribute to organizations' competitiveness in the medium and long term. Many business owners, directors, and managers have a false belief that QMSs provide immediate impact on competitiveness. Despite quality improvement efforts being sources of competitiveness, the essence of QMSs has been related to the impact on financial performance and customer satisfaction (Figure 1.1). These are the key points on which organizations should focus, since they add direct value to stakeholders.

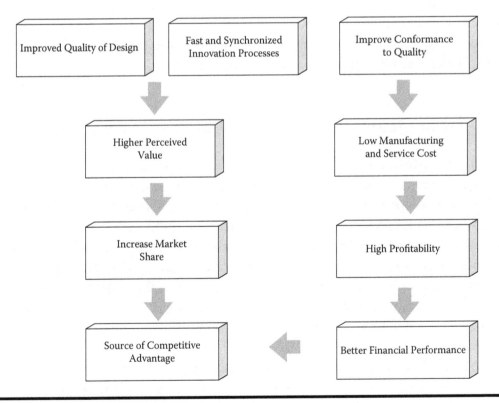

Figure 1.1   Impact of a QMS in financial performance, market share, and competitive advantage.

1.1.3 Innovation and Design

As consumer markets have evolved, customers have become more sophisticated, demanding better-quality products and services. Their perceptions
of quality have changed, and this fact increases the complexity of product
and service developments. This has resulted in efforts of product innovation,
putting pressure on all the supply chain to speed up production processes
to satisfy demands. For example, in the mobile telecommunications industry,
when the first mobiles phones were introduced in the 1980s, customers were
satisfied with basic functions to operate such devices. In those days, organizations complied with emerging legislations and quality standards, adjusting processes and the supply chain for that purpose. In the 21st century,
however, customers expect much more than basic functions. They prefer
devices with better displays, built-in entertainment, fast Internet connections,
productivity programs, accessories, and much more. To satisfy these customers' demands, organizations have been forced to adopt better-quality management practices. Thus, innovation and design are critical now, and this
requires an integrated approach to designing the quality and business management systems, based on the understanding of the organization's needs,
capabilities, business policies, and strategy direction.

1.2 From Quality Inspection to Business Excellence

Over the past decades, organizations have improved the quality of their products and services using a number of quality management initiatives (QMIs).
They have adopted these initiatives in response to increasing competition,
customers' demands, and technological changes. QMIs have played a major
role for these firms, helping them to increase productivity and achieve quality improvement goals. Among these initiatives, Total Quality Management
(TQM) represents a milestone in the development of the quality management
field. The definition provided by Porter and Tanner (1998) is a good approach
to describe what TQM means in a current organizational context:

> Total Quality Management is a business approach that focuses on
> improving the organization's effectiveness, efficiency and respon-
> siveness to customers' needs by actively involving people in pro-
> cess improvement activities.

This definition is accurate because it states the overall approach in a business context; it also pays attention to the original concept by focusing on improving operational performance based upon customers' needs. Before the TQM framework's introduction, many efforts had been made to integrate improvement activities within a single approach. Studying those efforts and early concepts of quality is important, not only for understanding the evolution of quality concepts into major frameworks, but also for adapting those frameworks as needed in fast-changing environments. It is therefore worthwhile to review how those efforts have helped to build the modern concepts of quality management systems.

During early days of manufacturing, products were accepted or rejected based on judgments as to whether those products were good or bad. The simple task of checking goods led production managers to create inspection activities in their daily operations. The motivation for improving quality grew, and soon statistical concepts helped enormously to improve quality control activities. It was perhaps early in the past century when the journey toward quality started with the introduction of Shewhart's and Deming's statistical charts, which represented the first tools to support quality control activities. Quality assurance was later supported by ISO standards, allowing the process of standardization. The focus of this approach was on prevention of poor quality and process improvement through the use of designed experiments, failure mode and effects analysis (FMEA), and standardization.

Later on, the principles and methods introduced by Deming, Crosby, Juran, Feigenbaum, Ishikawa, and Shingo came to constitute much of what today comprises the theory of quality management (QM). Then, the quality initiatives focused on process improvement, such as reengineering, Six Sigma, Lean principles, and updated ISO standards, among others, which complemented the quality management field. Beecroft (2004) suggests that there are "four major quality eras," as shown in Figure 1.2, basing his idea on the work by Garvin (1988), who suggested TQM and strategic quality management were the latest eras of quality. However, Garvin's (1988) perceptions of the quality eras were from the late 1980s, just a short time before the business excellence model was conceived (Conti, 2007), and following the introduction of the Malcolm Baldrige model in 1989. Since then, BEMs have played an important role in improvement activities, and have moved from the original concept of quality and TQM principles to a whole business-based approach. Excelling in business is the objective of a whole approach that evolved from inspection and statistical process control (SPC)

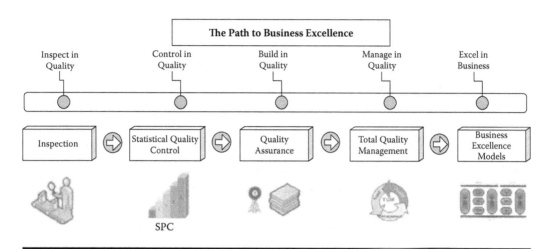

Figure 1.2   The path to business excellence.

basic concepts to a whole approach based on business criteria focused on key results and performance.

Originally, BEMs were based on and used the term *TQM*; then there was a shift of the term in a review of the framework in 1999 (Adebanjo, 2001). Since then, TQM has not been mentioned in the framework, as *business excellence* has become the common term to refer to the new quality era.

Figure 1.2 incorporates the new quality era, which refers to the era of BEMs, and which comprises areas of self-assessment, performance measurement, process improvement, business criteria and principles, etc. Those issues are discussed in forthcoming sections of this book.

1.3 A Systems Approach to Quality Management Systems

It is important to develop an understanding of how a business operates as a whole. The organization itself is considered to be a system, with a set of inputs and outputs and with interrelations between its elements. Business systems can be as complex as any other biological, physical, or mechanical system in nature. Business systems are composed of resources such as capital, knowledge, human resources, property, and facilities, among others. The approach to seeing an organization as a system helps to reduce complexity and understand the way the elements interact with each other. In this way, many organizations are divided by business units, divisions, departments, areas, products, and so on. Once the business system is understood and arranged in an optimal way, key and supporting processes should be identified to deploy best management practices. This can result in an optimal business array that produces and delivers products and services like an efficient engine. Within this context, quality management systems become a subsystem of the organization that should help to manage all issues related not only to the quality of products and services but also to a full business performance approach. Some key benefits of a business system approach are

- Reducing complexity of the whole business system by dividing it into subsystems
- Fully understanding how the business operates and why it operates in this way
- Identifying key and supporting processes that add value to stakeholders
- Optimizing and better allocating resources where/when required
- Modeling techniques that can be applied to predict future scenarios

The practical implications of having a business system approach are that it provides organizations with the ability to improve decision making at all levels in their business environments. This requires that organizations integrate a whole approach business management, performance measuring, quality management systems, and information systems aligned with strategic direction. The design of such systems should be carried out carefully with an engineering approach. It also requires at some point rethinking the way organizations operate to achieve optimal performance, and keeping continuous improvement of the business system.

1.3.1 Quality Management Systems and Business Strategy Alignment

QMSs and business strategy are some of the most widely discussed areas of knowledge in the business context. Both have provided methods and techniques for managing and improving the way organizations conduct their business. QMSs have given rise to some of the most popular models, methods, and quality tools, such as TQM, BEMs, ISO, Six Sigma, Lean approaches, and business process reengineering (BPR), among others. They are well documented in the literature. On the other hand, business strategy has provided theoretical foundations for extrapolating and understanding strategy concepts into the business arena (Mintzberg et al., 2000). Both concepts may sound distant, but actually under the business and organizational context they are closely related and require a clear understanding of their dynamic relationship (Beecroft, 1999).

Usually the lack of connection between QMSs and business strategy leads implementations to fail (Taylor and Wright, 2003). Sebastianelli and Tamimi (2003) found that this issue was the most significant factor that inhibits good TQM implementations, and that causes poor results in terms of the desired goals. Similarly, Ngai and Cheng (1997) argue that lack of vision and mission was the most significant barrier to implementing a TQM initiative. The problem has also been identified in other QMIs. Al-Mashari and Mohamed (1999) and Terziovski et al. (2003) argue that business process reengineering projects' lacking of alignment with business strategy has become a major barrier to success in BPR implementation.

The fact is that the lack of connection of QMSs with business strategy has significantly affected the success of implementations. The literature is full of misleading terms when referring to this issue. For this reason, it is very important that practitioners understand the concepts related to strategic planning, organizational objectives, strategic quality planning, and the business strategy as a whole. The problem encountered is that QMSs do not form part of the business strategy agenda or are not included in key objectives of strategic planning. The other side of the problem is that there is no strategic quality planning, that is, how organizations plan to mature the QMS in the medium and long term. The problem also concerns the models, methods, quality tools, capital, human resources, and time needed to reach business objectives. This issue is discussed in Chapter 6 when the development of the quality management system is proposed with the methodology of this book.

As a result of the lack of strategic direction and strategic quality planning, many quality management initiatives and the whole QMS fail in implementation or produce poor results. Another issue is also related to *management* and *leadership*, which are essential to deploy the QMS properly. Management is the only part of the quality management system where people can be the problem because management's decision making affects implementation and deployment. Managers need to be well trained to learn competencies and management abilities to successfully conduct projects. If the right people are not in management, then regardless of the good design of the QMS and the talented people we have, it will not be successful. Therefore, understanding strategic alignment, strategic quality planning, and getting the right people for the right quality management projects are essential to successfully design, build, and improve the quality management system. Strategic quality planning is addressed later in Chapter 5, providing practical issues in order to build the QMS.

1.4 Measuring QMS Performance

Measuring QMS performance is a relevant activity because it provides feedback and learning to organizations in relation to the effectiveness of the whole set of quality models, methods, and tools. Measuring QMS performance, however, is not an easy task, because it involves dealing with intangible variables and metrics that need to be accurately defined for proper interpretation. It is very often necessary to create frameworks to determine levels of success or failure to evaluate QMS implementations. The construction of such frameworks differs from organization to organization, depending on their objectives, industry, and reasons for implementing the *quality management system.*

When establishing specific metrics, they must reflect business performance while being accurate. Organizations usually set objectives for their areas but not specific metrics. For example, an objective is to increase financial performance by 8% for the next year. This sounds measurable but needs to be defined meaningfully. It is necessary to better define what *financial performance* means. It can be operational costs, increase in sales, reduction of inventory, or return of investments, among other key financial ratios. Ambiguous terms and inaccurate metrics are misleading and create inefficiency.

Since early implementation, it has been a tendency to measure QMS performance in terms of operational measures. Attempting to evaluate the performance of a QMS in several areas can be complex, as it is necessary to look at several parameters at the same time. It is more desirable to consider local measures of parts of the QMS (*quality management initiatives*) on a project-based approach than on complex evaluation frameworks. The metrics to measure have to be clear; measuring in terms of areas of financial performance, customer satisfaction, operational effectiveness, and market share is recommended. Most operational measures are frequently associated directly or indirectly with the organization's finances. It is no surprise that the first studies in QMS performance were based on financial performance and market value issues (Hendricks and Singhal, 2001; Eriksson and Garvare, 2005). The benefits have to be specific, and it is important to pay attention to decide when to measure the performance of the QMS (Taylor and Wright, 2003). If measures are taken too soon, the results are inconclusive; on the other hand, if measures are taken too late, time elapses and measures are affected.

Finally, *quality management systems* are not fully responsible for an organization's financial performance; this needs to be stated. Other internal and external organizational factors also affect financial results and overall performance. QMSs are just a part of the whole business system and, if deployed appropriately, can help to significantly improve business results. Thus, the results presented in the literature and by other organizations should be weighed and interpreted correctly, without assuming that fantastic or fair results are merely attributed to the implementation of a particular QMS.

1.5 Understanding the Way to Business Excellence

Developing QMSs and making them to work efficiently and successfully are a challenge for any organization. Many of them invest a considerable amount of human resources, capital, and time to build the right QMS. Very frequently those methods and tools are not adequate and are poorly deployed. In addition, in some cases QMSs are not aligned with strategic quality management and business strategies. This misalignment delivers inadequate implementations and pitfalls. To avoid it, the resources and capabilities should be assessed, and management should have a strategic quality plan to deploy and allocate resources to accomplish it. We show in Chapter 5 how to set the strategic quality plan.

1.5.1 Understanding the Vision and the Future

Management should focus on the future in order to direct the organization to the desired objectives, and the QMS is essential to support this action. The future is something that will probably present before we need it, and we have to be prepared to make any necessary changes that lead organizations to their objectives. In this context, making things with high-quality standards has to be a way of life and not a set of rules that people are obliged to do. A shared vision and values are therefore necessary to point out quality issues across the organization. Make sure that people understand and apply the core values at any time, and most importantly, build and spread a strong shared vision that leads all efforts to a single objective: achieve business excellence.

Once people understand where they are leading, it is easier to plan and allocate the resources to get to the desired objective. In this way, this book presents a practical way to achieve operational excellence, link it to business strategies, and the long-term decision making. This is a fundamental issue that management frequently fails to plan systematically and deploy with discipline. We need to warn that the road to business excellence is a never-ending process, full of challenges to overcome, but at the same time it is an exciting process that will pay off all efforts and resources invested.

1.6 Summary

This chapter has provided an overview of the QMS and focuses on the importance it has in fostering competitiveness, innovation, and providing high-standard quality products and services. The chapter has provided a review of the evolution of QMSs, from inspection to business excellence. In this way, it has analyzed QMSs from their first stages focused on inspection to contemporary business excellence models (BEMs) based on specific performance criteria. It has addressed the relevance of considering QMSs as systems integrated with a general business system approach, focused on business strategy, processes, customers, human capital, knowledge management, and IT. The chapter has brought to context in the first instance the relevance of understanding business strategy and QMSs, pointing out the importance of aligning strategic planning, strategic quality planning, and QMS design, implementation, and evaluation. It has argued the challenges of measuring QMS benefits in terms of financial performance, market share, competitiveness, and growth when designing or effectively implementing and applying QMSs. Some examples have been provided to support this point of view, and further readings about the topic are suggested to the practitioner to complement this section. Finally, the chapter closes setting a challenge for all kinds of professionals to build an effective QMS.

1.6.1 Key Points to Remember

- QMSs have to be able to produce high-quality products and services that eventually will foster innovation and competitiveness.
- Understand the evolution of quality management in order to identify current trends in QMSs along with business quality models, methods, and tools.
- Set a systems approach to integrate QMSs with the overall business system to maximize customer value and provide a strong basis for continuous improvement and change.

- Be sure to link business strategy with strategic quality planning and QMSs. This is a must to successfully implement your quality management initiatives.
- State in your proposals for business process improvements the impact that QMSs have in financial performance, operational cost reduction, efficiency, and productivity. Highlight the importance in the medium to long terms of achieving ongoing innovation, competitiveness, and market share.
- Share your vision of achieving a world-class organization with the others by carefully planning and considering the resources you need to get there. Review the suggested readings below, such as Collins (2001).

References

Adebanjo, D. (2001). TQM and business excellence: Is there really a conflict? *Measuring Business Excellence*, Vol. 5, No. 3, pp. 37–40.

Al-Mashari, M., and Mohamed, Z. (1999). BPR implementation process: An analysis of key success and failure factors. *Business Process Management Journal*, Vol. 5, No. 1, pp. 87–112.

Beecroft, G. D. (1999). The role of quality in strategic management. *Management Decision*, Vol. 37, No. 6, pp. 499–502.

Beecroft, G. D. (2004). Evolving quality improvement/implementation strategies. *ASQ's Annual Quality Congress Proceedings 2004, Quality Congress*, pp. 425–430.

Cobb, C. (2001). *From quality to business excellence: A systems approach to management.* American Society for Quality, Milwaukee.

Collins, J. (2001). *Good to great.* Harper Business, New York.

Conti, T. A. (2007). A history and review of the European quality award model. *The TQM Magazine*, Vol. 19, No. 2, pp. 112–128.

Eriksson, H., and Garvare, R. (2005). Organizational performance improvement through quality award process participation. *International Journal of Quality and Reliability Management*, Vol. 22, No. 9, pp. 894–912.

Garvin, D. A. (1988). *Managing quality.* The Free Press, Macmillan, New York.

Hendricks, K. B., and Singhal, V. R. (2001). Firm characteristics, total quality management, and financial performance. *Journal of Operations Management*, Vol. 19, No. 3, pp. 269–285.

Mintzberg, H., Ahlstrand, B., and Lampel, J. (2000). Strategy, blind men and the elephant. In Financial Times (Ed.), *Mastering strategy.* Financial Times-Prentice Hall, Hampshire.

Ngai, E. W. T., and Cheng, T. C. E. (1997). Identifying potential barriers to total quality management using principal component analysis and correspondence analysis. *International Journal of Quality and Reliability Management*, Vol. 14, No. 4, pp. 391–408.

Porter, L. J., and Tanner, S. J. (1998). *Assessing business excellence*. Butterworth-Heinemann, Woburn, MA.

Porter, M. (1998). *Competitive strategy: Techniques for analyzing industries and competitors*. Free Press, New York.

Sebastianelli, R., and Tamimi, N. (2003). Understanding the obstacles to TQM success. *Quality Management Journal*, Vol. 10, No. 3, pp. 45–56.

Taylor, W. A., and Wright, G. H. (2003). A longitudinal study of TQM implementation: Factors influencing success and failure. *International Journal of Management Science*, Vol. 31, No. 2, pp. 97–111.

Terziovski, M., Fitzpatrick, P., and O'Neill, P. (2003). Successful predictors of business process reengineering (BPR) in financial services. *International Journal of Production Economics*, Vol. 84, No. 1, pp. 35–50.

Further Suggested Reading

Conti, T. (2010). Systems thinking in quality management. *TQM Journal*, Vol. 22, No. 4, pp. 352–368.

Easton, G. S., and Jarell, S. L. (1999). The emerging academic research on the link between total quality management and corporate financial performance: A critical review. In Stahl, M. J. (Ed.), *Perspectives on total quality management*. Blackwell Publishers, in association with the American Society for Quality, Oxford.

Flynn, B. B., Flynn, E. J., Amundson, S. K., and Schroeder, R. G. (1999). Design quality and new product development. In Stahl, M. J. (Ed.), *Perspectives on total quality management*. Blackwell Publishers, in association with the American Society for Quality, Oxford.

Imler, K. (2005). *Get it right: A guide to strategic quality systems*. ASQ Quality Press, Milwaukee.

Link, A., and Scott, J. (2011). *Economic evaluation of the Baldrige Performance Excellence Program*, Planning Report 11-2. National Institute of Standards and Technology, U.S. Department of Commerce.

Toffler, A., and Toffler, H. (2006). *Revolutionary wealth: How it will be created and how it will change our lives*. Knopf, New York.

Chapter 2

Business Excellence Models

2.1 Introduction—QMSs and Business Models

A business model describes the way an organization develops, transforms, and delivers its products and services to the market. In other words, it describes the rationale and relationships required by an organization to operate in its industry. Understanding this is fundamental to determining the role of quality management systems (QMSs) and how they relate to the business model under a systems approach. Several elements should be considered in developing business models, namely, the customer segments, value propositions, distribution channels, customer relationships, revenue streams, key resources, key activities, partnerships, and cost of structure (Ostenwalder and Pigneur, 2010). In addition to these elements, a business should conduct an industry analysis through benchmarking, and ensure that it has the required resources and capabilities to operate. Understanding these elements may seem trivial, but very few individuals and companies deploy and correctly integrate the right business elements and tools to succeed.

Figure 2.1 shows the rationale for QMSs, the business model, and the overall business strategy. The core elements for operating as a system that delivers value to internal and external customers are based on human capital, key processes, information technology (IT), and knowledge management (KM). To be successfully integrated into a full business model, all of these elements require a structure, documentation, and effective management. *Customers* are the reason for any organization; without them, there is no reason to exist. Product development and innovation should be integrated to capture the voice of the customer (VOC) and produce high-value-added

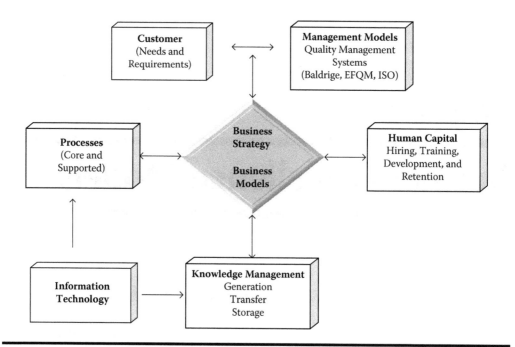

Figure 2.1   QMSs and the business model.

products that meet customers' expectations and lead to loyalty. This is per-haps one of the biggest challenges businesses face, and QMSs have to con-tribute to developing high-quality products/services and excel in customer service before, during, and after sales.

In this way, any organization that is serious about quality must have a sys-tematic approach to ensure customer satisfaction and create loyalty through a customer relationship management (CRM) system. A CRM system can be used to help manage and resolve customer complaints, deliver customer satisfaction surveys, and provide a system for collecting defective products or follow-up with corrective action. The system should also be able to provide competitive bench-marking, translate the VOC, and deploy focus groups to capture customers' needs and requirements for process development and innovation. We therefore recommend business intelligence to collect, analyze, and process all customer information through a systematic approach supported by a CRM system.

Human capital is a fundamental element of a full business system. People at all levels of the organization make most strategic and operational decisions and are fully responsible for business performance. Management personnel are also responsible for the development and implementation of a quality management system, and they have to ensure that they get the right people in the right positions. If management personnel are not in the right positions, regardless of their level of ability, commitment, and dedication, it

will result in poor QMS implementation and ultimately poor business performance. It is therefore compulsory to ensure that the best people are placed
in management positions by hiring and training them and retaining the best
talent for such positions. Similarly, people at the operational level must be
trained and retained since this reduces operational costs compared with the
cost incurred when a high level of personnel rotation exists. To address this
problem, Imler (2005) suggests the following:

- Hire, reward, and retain the right people.
- Make certain to get the right people for the right jobs.
- Retrain or get rid of people that do not contribute.
- Recognize that management could be the wrong people.

Thus, as a senior manager, you should focus on getting, keeping, and
providing professional development for the human capital through a systematic approach linked to human resource strategies.

Managing an organization as an entire system can be a complex job.
The identification of subsystems within the business model and key processes is a good strategy for simplifying this job. The notion of viewing
a business as a whole system is not a new approach; however, very few
people and companies truly have a systems-thinking approach when it
comes to understanding the cause–effect relationships that happen on a
daily basis. Most companies still work with "functional" areas that get lost
amid the hundreds of daily activities, and lose sight of the key processes
that affect business performance and value. Hence, there is a need to
focus on the things that add value to all stakeholders using a process-centered approach.

Core or value creation processes, in particular, are the activities that must
be well documented in QMSs in order to ensure the effectiveness of a business. However, identifying core processes may not be an easy task since
management may not have a good understanding of what actually adds
value to their activities and how these core processes are related to other
organizational structures, people, and technology. Every organization has its
own business units, divisions, and departments, all of which have different
requirements. The identification of core processes and their value can be
even more complicated if the management people do not have and share a
systems-thinking approach across the entire organization. In other words, it
is crucial to understand what actually adds value to the business system and
what does not. Therefore, organizations should make the effort to identify

their core processes, assign owners, and define the metrics and controls required to achieve the expected performance.

Supported processes, on the other hand, are those that will help ensure that core processes effectively and efficiently achieve an organization's requirements and objectives. Like core processes, they have to be well documented, with owners and specific metrics to track performance periodically. Chapter 3 discusses why process management is a key issue to consider and how it supports the proper development and deployment of quality management systems.

As an exercise, a set of activities that can be categorized as (1) value-added, (2) business value-added, and (3) non-value-added are shown in Table 2.1 so that we can see the value they add to an organization. This simple exercise helps to identify the activities that add or do not add value in single operations. When mapping a process, organizations should be able

Table 2.1   Value-Added and Non-Value-Added Activities

Activity	Value-Added	Business Value-Added	Non-Value-Added
Attending a weekly meeting with a project team		✓	
Reviewing and filtering e-mail lists			✓
Reporting performance to upper management		✓	
Planning an improvement program		✓	
Creating ISO documentation	✓	✓	
Building a best-practice database		✓	
Collecting information across departments to do your job			✓
Gaining multiple signatures/approvals to process information			✓
Assigning a tracking number to a complaint	✓		
Negotiating deliveries with suppliers	✓		
Communicating with your colleagues about a delay in a project			✓
Talking to your manager about your next promotion			✓
Getting training in leadership abilities		✓	

to identify and classify their activities as well as place them in logical order to make a process efficient. Let it be sufficient to say that this task of categorizing and prioritizing can significantly help to reduce waste, which has a direct effect in reducing operational costs.

Technology, particularly *information technology*, plays an important role in supporting core and supported processes as well as knowledge management efforts. By technology we mean all kinds of machines, software, hardware, industrial designs, patents, and special programs employed throughout the entire production chain, deliveries, and after-sales services that use scientific knowledge in a practical way. By information technology (IT), we mean all hardware—mobile devices, software, computer systems, and infrastructure and enterprise systems—that administers business information to support automation activities. Hammer and Champy (1993) argue that IT is a key element in successfully managing and automating processes from a reengineering approach. In the beginning of the dot.com era, IT was seen as a competitive advantage for outperforming rivals. In today's competitive business environment, the management of IT is essential to the survival of a business. This resource is even more critical for online and technology-based companies such as Amazon, Google, Microsoft, HP, and Apple, just to name a few. In fact, no company around the world, whether a manufacturing or service-based organization, can subsist without IT platforms that automate activities and manage business information at all levels.

Rarely do senior managers fully address or foresee the real problems when deploying IT projects (i.e., IT policies, legal use of information, investments, change management, technical feasibility, available technology, and training, among others). In many cases the results can be disappointing. IT alone will not solve any problem, nor will it automate and make business processes or an entire company more efficient. The key issue here is to understand the requirements and needs of organizations in administering their information at strategic and operational levels. Then, it is strongly suggested that a consulting team transfer all these needs and requirements into a cost–benefit and effective solution to administer business information. This requires a deep understanding of the systems-thinking and process-thinking approaches with the integration of technology and, most importantly, how it interacts with and supports the business model.

Knowledge management is also considered in this approach because, when deployed properly through structured programs, it can provide a framework for understanding and administering the way information is generated, stored, and transferred across the organization. The QMS, along with its models, methodologies, policies, tools, processes, procedures, etc., is itself a set of information that needs to be properly managed. The purpose is to have agile information systems that can provide business information at the right time to the right people when making business decisions. Whether this information is for market analysis, process improvement, product innovation, or business or financial performance, it must provide managers with a clear picture of the issue so that they can make the best decisions. Therefore, understanding knowledge management and how it can support quality management systems should be a priority for companies with medium- and high-quality maturity levels.

2.2 Business Excellence Models

Business excellence models (BEMs) are quality management frameworks based on organizational performance criteria that originated through the evolution of Total Quality Management (TQM) principles. BEMs have played a significant role in improving business among organizations, and these efforts are well documented with quality foundations that administer BEMs across regions and countries.* BEMs have witnessed an important evolution since their introduction in the late 1980s, not only in their business criteria but also in the way they are deployed and used. In this context, organizations have learned from the use and practice of these frameworks to apply the BEMs for several purposes. We have identified the following purposes: (1) award participation, (2) self-assessment, (3) business process improvement, (4) measurement systems, and (5) strategic planning (Rocha-Lona et al., 2008). Stating the specific role of the BEMs helps clarify objectives as well as determine the allocation of resources to a particular project improvement.

BEMs have been implemented to manage several organizations' categories to facilitate the assessment of their own business in terms of specific business criteria in their industry. Initially, those categories were better suited for large public and private organizations. However, the necessity to include and expand BEMs to most industrial sectors encouraged quality foundations to develop frameworks for other types of organizations. The introduction of new categories to frameworks, such as healthcare, nonprofit,

---

* See the websites of the European Foundation for Quality Management, National Institute of Science and Technology, Japanese Institute of Scientists and Engineers, and Canadian Quality Assurance Institute, among others.

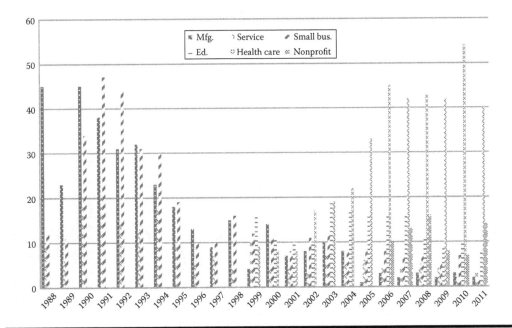

Figure 2.2   Baldrige applications from 1988 to 2011. (Data from NIST, Baldrige Award Recipients, Contacts and Profiles, National Institute of Standards and Technology, 2012, available at http://www.baldrige.nist.gov/Contacts_Profiles.htm [accessed May 20, 2012].)

education, and medium and small organizations, profoundly helped to increase the use of BEMs. Figure 2.2 shows, for instance, that applications for the Malcolm Baldrige National Quality Award (MBNQA) have increased in the last years after having suffered a setback in 1997. This increase may directly correspond to the introduction of new categories in the late 1990s. Thus, it is reasonable to think that the use of BEMs may continue to grow as the quality foundations continue innovating the frameworks for industrial sectors or specific products and services.

Table 2.2 shows some of the most popular BEMs along with their *business model criteria*. The models are categorized based on region, industry, type of organization, and their business criteria. These criteria have evolved and have been adapted according to organizations' needs, and they usually change yearly or every two years, depending on the decisions made by reviewing committees. The recommendation is that organizations look for the most recent business model criteria when deciding to implement one of these frameworks. This ensures that an organization has updated criteria that address the current issues of the business, particularly those related to industry regulations. For a more comprehensive list of BEMs in several countries, see, for instance, Mohammad and Mann (2010).

Table 2.2 Business Excellence Models—Industries and Model Criteria

Business Excellence Model	Country/Region (Primary Partners)	Website	Category/Sector	Model Criteria
Malcolm Baldrige	United States	http://www.nist.gov/baldrige/	Manufacturing Service Education Healthcare Nonprofit	1. Leadership 2. Strategic planning 3. Customer focus 4. Measurement, analysis, and knowledge management 5. Workforce 6. Operation focus 7. Results
EFQM model	UK, Germany, some other countries in Europe	http://www.efqm.org	All types of organizations and industries	1. Leadership 2. People 3. Strategy 4. Partnerships and resources 5. Processes, products, and services 6. People results 7. Customer results 8. Society results 9. Key results
Deming Prize	Japan	http://www.juse.or.jp	Individuals (in Japan) Individuals (overseas) Organizations (in Japan) Organizations (overseas) All kinds of organizations, including small and medium-sized enterprises (SMEs), business unit divisions, and headquarters	Specifically, the Deming Prize Committee set the following criteria: 1. Customer-oriented business objectives and strategies are established in a positive manner according to the management philosophy, type of industry, business scale, and business environment with the clear management belief. 2. TQM has been implemented properly to achieve business objectives and strategies as mentioned in item 1. 3. The business objectives and strategies in item 1 have been achieving effects as an outcome of item 2.

Award	Country	Website	Sectors/Categories	Model Criteria
Canadian Awards for Excellence	Canada	http://www.nqi.ca	Innovation and wellness (formerly integrated quality and healthy workplace) Quality (private and public sectors) Healthy workplace Healthy workplace for SMEs Education (K–12) Quality and customer service for SMEs Community building Projects Senior-wise Mental health at work	Several model criteria depending on the award
Shingo Prize	Mexico, United States, Canada	http://www.shingoprizemexico.org	Manufacturing organizations	Cultural enablers Continuous process improvement Enterprise alignment Results
Prêmio Nacional da Qualidade	Brazil	http://www.fnq.org.br	Open to public and private organizations	1. Leadership 2. Strategies and plans 3. Customers 4. Society 5. Information and knowledge 6. People 7. Processes 8. Results

2.3 Evolution of BEMs

Most BEMs have evolved through time in response to internal and external changes produced by social, economic, and technological factors. The Baldrige model has perhaps evolved more consistently than any other model. The American Society for Quality (ASQ) and the National Institute for Science and Technology (NIST) have the responsibility of updating it every year. The evolution of this framework is remarkable in terms of the business criteria that have consistently evolved to address most of America's business needs, technological issues, and even extreme social events (Rocha-Lona, 2012). Since the Baldrige model was one of the first BEMs used around the world, many governments and organizations have used it as a standard by which to develop their own quality frameworks. Companies started using BEMs for self-assessment, and then moved quickly from using BEMs for award participation to a more holistic approach (Ahmed et al., 2003). So, we explain the big shifts that BEMs have undergone in their evolution. This is illustrated in Figure 2.3.

Following the focus of BEMs on award participation, organizations have used *self-assessment* to obtain a "picture" of their business processes on a regular basis and identify areas in need of improvement. Conducting an assessment and interpreting its results requires discipline and objectivity. To address objectivity, some organizations use external services to ensure that the outcomes of this process accurately reflect the state of the business. Thus, focus on self-assessment is widely accepted as a systematic and regular view of an organization's activities.

BEMs have also been used to coordinate *improvement programs* because organizations employ the self-assessment process outcomes for quality improvement purposes. In this way, the identification of improvement areas for quality purposes is one of the main benefits of using self-assessment. The information, in the form of reports, is passed on to top management for its analysis and further use; however, in many cases there is no way to know about or track further actions. The process ends with these reports; consequently, it is the ability of top management to decide *what* areas are

Figure 2.3   Use of business excellence models.

priorities and *how* to improve those areas through specific improvement programs. The success of this process may be limited to the correct interpretation of top management and the available guidance in effectively using the self-assessment outcomes.

After using BEMs to identify improvement areas, organizations recognized the suitability of the models for *measuring organizational performance*. This recognition was derived by performing self-assessment and measuring key areas of the business. BEMs were not originally designed to measure organizational performance; however, they present a broader view of performance, addressing many areas not dealt with through other approaches (Kennerley and Neely, 2002). Consequently, an interest in employing BEMs for developing performance measurement systems has increased.

Finally, *strategic planning and decision making* is the last role that BEMs have adopted. However, caution should be exercised since BEMs have not been fully applied to the area of business performance measurement, and there is still little evidence regarding the true impact of models in developing and deploying strategic planning. Self-assessment outcomes should be able to support business plans at strategic and operational levels, and some business model criteria are more suitable for supporting strategic planning than others. For example, organizational effectiveness and customer results can serve as effective criteria upon which an organization can base its strategic analysis at a given point in time. Other organizations find market and financial results valuable for setting strategic objectives and future plans (Rocha-Lona, 2012). This tells us that organizations do not have the same priorities in selecting the criteria that best support their strategic planning processes. This will vary depending on ongoing strategies, objectives, quality improvements, plans, and the maturity level of the organization administering the self-assessment process. When planning which BEMs to adopt, it is essential to invest time in order to select the right BEM's role that matches business objectives. This will ensure the right focus for the model and the relevant resources and planning activities for the deployment process.

2.4 Comparison of QMSs

In terms of their objectives, TQM, International Organization for Standardization (ISO) standards, and BEMs are similar and aim to be quality management systems that lead organizations to become world-class (Table 2.3). BEM and ISO standards were structured based upon

Table 2.3 Comparison of QMSs

Category	TQM	ISO 9000	Malcolm Baldrige	EFQM
Objective	To help organizations produce value-for-money products and services that meet customer expectations (Dale, 2003).	To ensure that products and services are reliable and of good quality according to international standards.	To improve the competitiveness and performance of U.S. organizations.	To be the driving force for sustainable excellence in organizations in Europe (EFQM, 2010).
Definition	"Total Quality Management is a business approach that focuses on improving the organization's effectiveness, efficiency, and responsiveness to customers' needs by actively involving people in process improvement activities" (Porter and Tanner, 1998).	The ISO 9000 family of standards is a set of norms that addresses several issues regarding quality management. Among the areas are requirements for the QMS, concepts and definitions, customer satisfaction, quality plans, and measurement systems.	The Baldrige model provides a systems perspective for understanding performance management.	The EFQM Excellence Model is a nonprescriptive framework based on nine criteria. Five of these are "enablers" and four are "results." The enabler criteria cover what an organization does. The results criteria cover what an organization achieves. Results are caused by enablers and enablers are improved using feedback from results.
Type	Set of principles supported by a range of methods and tools.	Set of standards and specific criteria to be covered.	A systematic approach of the whole business, process centered and focused on business performance.	Systematic approach of the whole business, originally encouraging self-assessment. Strongly results oriented.
Principles	Customer focus Leadership Involvement of people Process approach System approach to management Continuous improvement Factual approach to decision making	Customer focus Leadership Involvement of people Process approach System approach to management Continuous improvement Factual approach to decision making		Based on fundamental concepts and nine criteria. Fundamental concepts: results orientation, customer focus, leadership and constancy of purposes, management by process and facts, people development and involvement, continuous learning, improvement and innovation, partnership development, and corporate social responsibility.

Beneficial supplier relationships (Dale, 2003)	Mutual beneficial decision making		Nine criteria: leadership, people, policy and strategy, partners and resources, processes, people results, customer results, society results, and key performance indicators.
Focus System based, Customer based, Process based	Process based	System based, Customer based, Process based, Results oriented	System based, Customer based, Process based, Results oriented
Methods and tools Statistical process control (SPC), Plan, do, study, act (PDSA) (Beecroft, 2004)		Self-assessment	Results, approach, deployment, assessment, and review (RADAR)
Drawbacks Prescriptive, with a set of methods and tools to support initiatives. TQM is conceptual and philosophical (Salegna and Fazel, 2000). Its lack of structure and overall approach makes it difficult for managers to interpret (Adebanjo, 2001).	Low flexibility in process, Does not encourage innovation processes, Expensive to deploy, No warranty exists for optimal performance	Nonprescriptive and offers overall guidelines to conduct self-assessment. Needs to avoid evolving into a mere point scoring system (Wang and Ahmed, 2001).	Nonprescriptive and offers overall guidelines to conduct self-assessment. Those activities are underpinned by so-called fundamental concepts (Have et al., 2003). Needs to avoid evolving into a mere point scoring system (Wang and Ahmed, 2001).

TQM principles, and they share some similar tools and techniques. However, while BEMs (such as the Baldrige and European Foundation for Quality Management (EFQM) models) have been tailored to organizations by region and "organization categories" (i.e., large organizations, small and medium organizations, educational, healthcare, nonprofit), TQM has remained open to most organizations. ISO standards, on the other hand, are specific and currently provide a wide range of standards for several industries.

BEMs and ISO standards are results oriented and process based, while TQM is project based. In terms of their definitions, TQM is still ambiguous and has several meanings and interpretations. In contrast, BEM and ISO are better defined as overall frameworks that attempt to look at a whole business by identifying areas of improvement and self-assessment, and by providing results section guidelines for financial and nonfinancial performance. In terms of the methods and techniques employed in making improvements, TQM provides a wide range of tools that include quality circles, statistical process control (SPC), and quality function deployment, among many others. BEMs provide scoring systems that leave managers the option of choosing their techniques and methods. Finally, ISO standards require adherence to several business criteria to ensure compliance with specific industry regulations.

Some of the main drawbacks of TQM are that the fretwork is conceptual and philosophical, leaving the initiatives to the correct interpretation and good judgment of managers. Furthermore, TQM lacks clear definitions and flexibility, making it difficult to define specific improvement programs and adapt them in the short and medium terms. Additionally, the wide acceptance of BEMs and ISO standards has slowed down the attention to and use of TQM, along with its tools and techniques (Adebanjo, 2001). Based on this, we can say that TQM, business excellence models, and ISO standards are different approaches that share some common objectives. BEMs represent the evolution of TQM principles, which have been adapted to a more process- and results-oriented approach. Hence, for the purpose of this book, we recommend focusing on BEMs as a general umbrella for deploying the quality management system.

2.5 Quality Management Standards

Quality management standards are the requirements and criteria organizations must meet to participate in regulated industries. When planning QMSs, it is essential to consider all standards and requirements so as to comply with the industry regulations in which an organization operates. Failure to comply with such standards could result in losing customer contracts and incurring government/agency fines. In the worst scenarios, companies are forced to close operations temporarily or permanently depending on the severity of the nonconformances. Thus, those specific needs regarding compliance have to be integrated into a strategic quality plan. Since the aim of this book is to provide some guidance for building QMSs, we will focus on the ISO 9000 series of standards. However, as mentioned previously, check the required standards for your specific organization's industry and follow its guidelines.

ISO is the International Organization for Standardization, founded in 1947. Since then, it has published more than 19,000 international standards for many industries related to technology and business. For the purpose of this text, we will focus on standards related to the administration of quality management systems. The ISO 9000 family addresses various aspects of quality management, which can be seen in Table 2.4.

Table 2.4   Main ISO Series of Standards for QMSs

Norm	Description
ISO 9000:2005	Focus on concepts and language
ISO 9001:2008	Sets out the requirements for a QMS
ISO 9004:2009	Focus on how to make the QMS more efficient and effective
ISO 20001:2007	QM—customer satisfaction—guidelines for codes of conduct of organizations
ISO 10002:2004	QM—customer satisfaction—guidelines for complaints handling in organizations
ISO 10003:2007	QM—customer satisfaction—guidelines for dispute resolution in external organizations
ISO 10005:2005	QMS—guidelines for quality plans
ISO 10006:2003	QMS—guidelines for quality management in projects
ISO 10007:2003	QMS—guidelines for configuration management
ISO 10012:3003	Measurement management systems—requirements for measurement processes and equipment

The ISO series of standards are some of the most widely used quality standards that have helped organizations with conformance to the quality standards of their products and services. Some of the main benefits that can be achieved with ISO quality standards are that activities are documented, improvizations are eliminated, and the quality of goods is uniform and in some way ensured. Additionally, customers tend to trust certified organizations rather than those that do not have certificates. Finally, the presence of ISO quality standards serves as a strong base for continuous improvement. Hence, there is a general benefit for customers, suppliers, employees, shareholders, and the community. On the other hand, some drawbacks include lack of flexibility in processes, process certification is expensive in most cases, the high quality of products and services is not completely assured, the excessive documentation required leads to bureaucracy, and there is no warranty to ensure optimal performance. These are some of the highlights that have been reported when quality management systems have been implemented based on ISO norms. However, despite the drawbacks, the ISO quality standards are still some of the most widely used norms for the regulation of many industries. In Table 2.5 we provide the main principles of the ISO 9000 standards, which are translated into the core benefits that can be achieved through the proper deployment of this QMS.

The real issue is that many organizations have been forced to adopt norms not as a quality management system, but as way to conform to certain industrial standards or comply with other companies in the supply chain. This leads to a very limited use of norms, and it fails to provide a real impact to the QMS in the medium and long terms from a strategic, continuous improvement standpoint. Therefore, when planning a QMS, organizations must ensure that the implementation of the ISO standard is part of the strategic quality plan (see Chapter 5).

2.6 Leading to an Integration of Management Standards

There are many specific ISO 9000 variations that combine criteria from the norm and industry-regulated requirements. Those requirements have evolved and continue to evolve over time; therefore, it is vital to find out whether an organization's QMS has to cover those requirements in order to adapt a specific norm. For example, the following are norms that might be considered:

Table 2.5   ISO 9000 Standards Management Principles

Management Principle	Explanation	Key Benefits	Business Excellence Model's Equivalent	Interpretation
Customer focus	Organizations depend on their customers and therefore should understand current and future customer needs, meet customer requirements, and strive to exceed customer expectations.	Increased revenue and market share obtained through flexible and fast responses to market opportunities Increased effectiveness in the use of the organization's resources to enhance customer satisfaction Improved customer loyalty leading to repeat business	Customer focus	Customers are the reason for any organization; without them, there is no reason to exist. Product development and innovation should be integrated to capture their voice and produce high-value-added products that meet customer expectations and will lead to loyalty.
Leadership	Leaders establish unity of purpose and direction of the organization. They should create and maintain the internal environment in which people can become fully involved in achieving the organization's objectives.	People will understand and be motivated toward the organization's goals and objectives Activities are evaluated, aligned, and implemented in a unified way Miscommunication between levels of an organization will be minimized	Leadership	Leadership is the most important ability for managers to have.
Involvement of people	People at all levels are the essence of an organization and their full involvement enables their abilities to be used for the organization's benefit.	Motivated, committed, and involved people within the organization Innovation and creativity in furthering the organization's objectives People being accountable for their own performance People eager to participate in and contribute to continual improvement	People workforce	Organizations are managed by people. They are a fundamental element of the business system. Ultimately, they make most strategic and operational decisions and are responsible for business performance.

Table 2.5 (Continued) ISO 9000 Standards Management Principles

Management Principle	Explanation	Key Benefits	Business Excellence Model's Equivalent	Interpretation
Process approach	A desired result is achieved more efficiently when activities and related resources are managed as a process.	Lower costs and shorter cycle times through effective use of resources Improved, consistent, and predictable results Focused and prioritized improvement opportunities	Process focus; value-added and business value-added processes	The process approach provides an effective way of managing organizations.
Systems approach to management	Identifying, understanding, and managing interrelated processes as a system contributes to the organization's effectiveness and efficiency in achieving its objectives.	Integration and alignment of the processes that will best achieve the desired results Ability to focus effort on the key processes Providing confidence to interested parties as to the consistency, effectiveness, and efficiency of the organization	Systems perspective	A systems perspective helps organizations understand relationships among processes and make them efficient and effective.
Continuous improvement	Continuous improvement of the organization's overall performance should be a permanent objective of an organization.	Performance advantage through improved organizational capabilities Alignment of improvement activities at all levels to an organization's strategic intent Flexibility to react quickly to opportunities		

Factual decision making	Effective decisions are based on the analysis of data and information.	Informed decisions An increased ability to demonstrate the effectiveness of past decisions through reference to factual records Increased ability to review, challenge, and change opinions and decisions
Mutually beneficial supplier relationships	An organization and its suppliers are interdependent, and a mutually beneficial relationship enhances the ability of both to create value.	Increased ability to create value for both parties Flexibility and speed of joint responses to changing market or customer needs and expectations Optimization of costs and resources

Source: ISO, *ISO Quality Management Principles*, ISO Central Secretariat, Geneva, 2012, available at http://www.iso.org.

- QS-9000 is a norm that was adopted in the automotive industry and has now evolved to TS-16949, which was developed by the International Automotive Task Force (IATF). The norm defines the requirements to design, develop, produce, and install devices for the automotive industry.
- AS-9100, developed by the Society of Automotive Engineers (SAE International), comprises the requirements for the aerospace industry. It also aligns with ISO 9001:2008.
- TL-9000 was developed by QuEST and comprises the supply chain quality requirements for the telecommunications industry.

There also many standards apart from ISO 9000:2005, such as ISO 14001:2004 for environmental management, ISO 22000 for food and safety, ISO 28000 for supply chain security, and ISO/IEC 27001 for information security. Many companies are now adopting Integrated Management System Standards (IMSS) in order to comply with several requirements for their regulated industries. In addition, there are other integrations that can be achieved with current process-based approaches such as Lean and Six Sigma. This has gained particular attention since organizations have to cover multiple aspects, levels, functions, and expectations for the quality management system and internal and external stakeholders. Therefore, the challenge is to set up an IMSS that aims for an effective and efficient way of meeting the requirements of the business system and covers multiple objectives related to industry regulations, customers, investors, government agencies, and the organization's own business quality management philosophy.

Deciding which QMS to implement is a highly strategic issue that must be addressed in formal strategic planning activities, and it is essential to understand implications in terms of resources, requirements, and business needs. This issue is addressed in Chapter 5, which covers how to establish a strategic quality plan and provides helpful tips for decision making.

2.7 Summary

This chapter presents the quality management system from a system and business perspective and integrates key areas such as customers, human capital, business processes, IT, and knowledge management. The chapter also discusses the evolution of BEMs from the award participation approach to the integration of self-assessment with strategic planning and decision making in order to make the best decision to select the right approach. It also presents a comparison of the main QMSs, such as BEMS, TQM, and ISO standards, to point out the main advantages and drawbacks of these QMSs. In addition, it briefly describes the ISO 9000 series of standards along with the specific industry-regulated standards. Finally, the chapter proposes the integration of all standards into a single management framework (IMSS) in order to comply with several industry regulations and internal and external stakeholders. It closes with a commentary regarding the importance of the selection of the right QMS based on what best suits an organization's needs, requirements, and resources.

2.7.1 Key Points to Remember

- Make sure your organization has a system approach to management and that the key elements are fully integrated with a business strategy.
- Understand the nature and potential benefits of deploying BEMs with specific purposes (i.e., award participation, process improvement, performance measurement system). This is a key factor in building the fundamentals of a QMS.
- Support the right selection of the QMS at the business strategy level by understanding the advantages and drawbacks of main QMSs, such as the TQM, BEM, and ISO 9000 series of standards. Strongly argue for your selection based on your organization's needs, requirements, and resources.
- Integrate, when possible, all your industry regulation requirements with an Integrated Management System Standard (IMSS). This will simplify all tasks for process certifications and make process management efficient.
- Make sure to select the right quality management model based on reliable business information and the aim of satisfying all stakeholders.

References

Adebanjo, D. (2001). TQM and business excellence: Is there really a conflict? *Measuring Business Excellence*, Vol. 5, No. 3, pp. 37–40.

Ahmed, A. M., Yang, J. B., and Dale, B. G. (2003). Self-assessment methodology: The route to business excellence. *Quality Management Journal*, Vol. 10, No. 1, pp. 43–57.

Beecroft, G. D. (2004). Evolving quality improvement/implementation strategies. ASQ's *Annual Quality Congress Proceedings 2004, Quality Congress*, pp. 425–430.

Dale, B. G. (2003). *Managing quality*. Blackwell Publishing, Oxford.

EFQM. (2010). *Overview EFQM 2010*. European Foundation for Quality Management.

Hammer, M., and Champy, J. (1993). *Reengineering the corporation: A manifesto for business revolution*. Harper Business, New York.

Have, S. T., Have, W. T., Stevens, F., Elst, M. V. D., and Pol-Coyne, F. (2003). *Key management models*. FT Prentice Hall, Glasgow, UK.

Imler, K. (2005). *Get it right: A guide to strategic quality systems*. ASQ Quality Press, Milwaukee.

ISO. (2012). *ISO quality management principles*. ISO Central Secretariat, Geneva. Available at www.iso.org.

Kennerley, M., and Neely, A. (2002). Performance measurement frameworks: A review. In Neely, A. (Ed.), *Business performance measurement*. Cambridge University Press, Cambridge.

Mohammad, M., and Mann, R. (2010). National Quality/Business Excellence Awards in different countries. Available at http://www.nist.gov/baldrige/community/upload/National_Quality_Business_Excellence_Awards_in_Different_Countries.xls (accessed September 23, 2012).

NIST. (2012). Baldrige award recipients, contacts and profiles. National Institute of Standards and Technology. Available at http://www.baldrige.nist.gov/Contacts_Profiles.htm (accessed May 20, 2012).

Ostenwalder, A., and Pigneur, Y. (2010). *Business model generation*. John Wiley & Sons, Hoboken, NJ.

Porter, L. J., and Tanner, S. J. (1998). *Assessing business excellence*. Butterworth-Heinemann, Woburn, MA.

Rocha-Lona, L. (2012). *Business excellence models and strategic planning*. LAP LAMBERT Academic Publishing, Germany.

Rocha-Lona, L., Eldridge, S., Barber, K. D., and Garza-Reyes, J. A. (2008). Using business excellence models for supporting decision making in strategic planning and business improvements. Proceedings of the 18th International Conference on Flexible Automation and Intelligent Manufacturing (FAIM), Skövde, Sweden, June 30–July 2.

Salegna, G., and Fazel, F. (2000). Obstacles to implementing quality. *Quality Progress*, Vol. 33, No. 7, pp. 53–57.

Wang, C. L., and Ahmed, P. K. (2001). Energising the organisation: A new agenda for business excellence. *Measuring Business Excellence*, Vol. 5, No. 4, pp. 22–27.

Further Suggested Readings

Conti, T. (2010). Systems thinking in quality management. *TQM Journal*, Vol. 22, No. 4, pp. 352–368.

ISO. (2008). *The integrated use of management standards*. International Organization for Standardization, Geneva.

Jonker, J., and Karapetrovic, S. (2004). Systems thinking for the integration of management systems. *Business Process Management Journal*, Vol. 10, No. 6, pp. 608–615.

López-Fresno, P. (2010). Implementation of an integrated management system in an airline: A case study. *TQM Journal*, Vol. 22, No. 6, pp. 629–647.

Chapter 3

Process Management

3.1 Introduction

The previous chapter discussed various business excellence models, quality management standards, quality methods, and tools. This chapter focuses on the significance of managing business processes in the overall performance improvement of an organization. The chapter starts with an emphasis on the need for the efficient and effective process management of organizations. A brief definition of process management is also provided. We have emphasized that managing quality within the organizations is very much dependent on the way the organizations manage their processes, and together they influence their overall performance.

The chapter also acknowledges the role of information technology (IT) in managing business processes and urges organizations to build IT competence. In order to do this they need to be familiar with their capacities and also well aware of the limitations of computer technology and its impact. If they fail to do this, then IT competence is hard to build up. The chapter then puts an emphasis on identifying core processes and argues that core business processes create real value in the organization. The chapter also briefly explains the role of value stream mapping (VSM) in identifying value-added and non-value-added activities in organizational processes. We conclude this chapter by suggesting that organizations need to have a well-defined process improvement agenda that merits good execution.

3.2 Managing by Processes

In the global competitive environment establishing a quality management system (QMS) has become a necessity. Achieving business excellence through quality improvement by following various methods and tools is one of the priorities for any type of organization. In addition, a changing competitive environment has forced organizations to evaluate carefully their competitive position in the industry, seek ways of building competitive advantage, and defend against the possible threats imposed by their rivals. Organizations therefore need to focus on, and evaluate, their external and internal environment prior to planning their strategy. On one hand, they need to evaluate their market position, product line, service quality, and customer satisfaction, while on the other hand, they need to identify their resource strengths and learn to execute activities more efficiently to gain competitive edge. The intense competitive rivalry has also left no other option for the successful organizations to afford any internal inconsistencies and inefficiencies. Thus, a right balance between internal efficiency and external effectiveness is required, which points toward the requirement of a well-designed business process. An efficient and effective management of processes is vital for the sustained performance of organizations. The mismanagement of processes can lead to significant losses to organizations in the form of unnecessary costs, poor quality, poor operational efficiency, and poor performance. Therefore, organizations must ensure that their processes are well managed.

3.2.1 Defining Processes

The management of processes is an essential element of the quality management system (QMS), as successful process management is vital for achieving goals of operational efficiency and quality improvement. Process management is often referred to as an activity or set of related activities that accomplishes a specific organizational goal, but it is also about planning and monitoring the performance of a process with the ultimate goal of profitably meeting customer expectations and requirements. In very simple terms, a process can be defined as the steps and decisions involved in accomplishing a task. The notion of process management is also very closely related to the principles of quality improvement, i.e., define, measure, analyze, improve, and control. Laguna and Marklund (2004), following the core principles of a successful process management proposed by Melan (1993), have divided it into three phases:

- Phase I: The initialization phase defines the entry and exit points of the processes by appointing a process manager. Thus, this phase involves assigning process ownership and analyzing process boundaries and interfaces.
- Phase II: The definition phase involves a thorough understanding of the process flow, activities, and facilitating communication among those involved in this process within the organization.
- Phase III: The third phase is the control phase, which aims at controlling the process and providing feedback to the people involved.

Being central to the transformation model, input-process-output of any manufacturing or service activity, process management always draws key attention from the business and operational managers. Process-oriented design is well established in practice and has been a major topic of discussion since the late 1980s, when organizations were referring to it as *business process management* (BPM) or *business process reengineering* (BPR). Nonetheless, many organizations often struggle to understand their own processes unless the management has implemented a well-established QMS to monitor their quality and processes. This lack of visibility in relation to processes poses significant problems for the management team in identifying the root cause of problems, further resulting in the deterioration of quality levels, reduced operational performance, and increased costs. A process is also conceptualized as the transformation from the product development stage to the final product, whereas business process reengineering focuses on the whole process. The concept of business process reengineering was a step further from the simple process management that was aimed at invoking fundamental rethinking and a thorough redesigning of business processes to obtain striking and sustained improvements in quality, cost, service, lead time, outcomes, flexibility, and innovation (Gunasekaran and Nath, 1997). Improvement in processes can significantly improve the performance of an organization. Thus, we would emphasize that organizations need to realize the importance of a proper understanding of processes and continuously seek ways through which processes can be managed more efficiently.

3.2.2 Importance of Process Management

In an ideal world, top management is responsible for drafting the vision and planning the strategy of the organization; however, in reality they are seldom involved in brief process planning and execution. Process management is primarily dealt with by the operational or functional-level managers. We do not mean to say that top management has no role in the process management and execution, but rather we want to emphasize that this is better looked after by the middle management, particularly managers who are responsible for looking after the processes within their departments. Managing business processes involves the identification and definition of processes, instituting responsibilities, evaluating performances, and exploring opportunities for further improvement. Therefore, the notion behind efficient process management is to improve the organization's work flow and make that organization capable of adjusting to the uncertain environment. There is plenty of evidence from companies around the world that highlights the significance of process management, such as software, manufacturing, or service companies that are successful following efficient process management practices.

Managing quality within the organizations is very much dependent on the way the organizations manage their processes, and together they influence their overall performance. Moreover, management of the end-to-end processes is an ongoing requirement if a company is to meet its customer requirements (Kumar et al., 2008). Often processes that involve complex routine work involving many people pose significant challenges to the management team, and therefore it is essential for the management team to understand, analyze, and continuously look for ways to improve the processes. To have a better understanding of processes, organizations first need to draw a process map/chart to increase their clarity on how different processes are interconnected. This understanding not only helps organizations to visualize their processes but also assists them in identifying the root cause of problems that are centered on the mismanagement of processes. The essence of managing processes is to indentify the best means of performing tasks meaningfully, effectively, and efficiently. Thus, we would like to stress that by improving processes organizations can improve their quality levels, complementing the QMS and leading to overall performance improvement. In the next section, we discuss the role of information technology (IT) in managing processes.

3.3 Role of Information Technology (IT)

Information technology (IT) is often referred to as the technology of coding, sensing, transmitting, translating, and transforming information. In the last couple of decades, with a rapid advancement in the technological arena, IT has become central to the modern organization's survival and growth. Regardless of the size of an organization, i.e., whether it is a small company or a large multinational company, all of them now rely on information technology in some way or another in their daily business practices. Despite the growing significance of IT in business process performance, many organizations still rely on the capabilities and performance of the team responsible for driving these processes. Slowly, however, organizations have started to rely more on IT to manage their business processes.

IT plays a multidimensional role in processing data, information gathering, storing collected materials, accumulating knowledge, and expediting communication (Chan, 2000). Further, the new advancements in IT, such as image processing and expert systems, can help organizations to reduce their non-value-added activities. The growing use of Enterprise Resource Planning (ERP) systems, customer relationship management (CRM) systems, management information systems (MISs), decision support systems (DSSs), Transmission Control Protocol/Internet Protocol (TCP/IP)-based modeling, etc., further reveals the increasing importance of IT in the modern competitive arena. Research evidence has shown that organizations failing to adopt IT systems are far less successful than their counterparts who have a well-established IT system embedded across their departments and are using IT systems effectively and efficiently (Bharadwaj, 2000). IT has a major role to play, particularly when we are discussing the significance of efficient and effective process management. These days IT is no longer seen as a supporting player, but rather has emerged as a key player in business processes—creating new needs, causing new product development, and commanding new procedures. Although IT is a key player, its implementation is not straightforward. Research evidence has shown that for organizations to be successful, they need to adopt IT as a part of their system or cluster of mutually reinforcing organizational changes, thus placing an emphasis on the issue that investment in information technology complements changes in other aspects of the organization. Hence, during IT implementation organizations have to overcome the challenges imposed by the need for the organizational change.

3.3.1 Developing IT Competence

Realizing the intense competitive environment, organizations need to develop IT competence to counter the threats imposed by their rivals. Many organizations have gained a competitive advantage using the power of IT.

Whether we look at the examples of Amazon.com, eBay, Wal-Mart, or Dell, all of these organizations have used the disruptive power of IT to break the rules and gain a significant competitive advantage in their relative field. Certainly IT can be a source of competitive advantage, but if they want to build IT competence, organizations need to be familiar with their capacities and be well aware of the limitations and impact of computer technology (Konar et al., 1986). This notion is also supported by the fact that for the successful implementation of IT, organizations need to make sure their structure is well matched with their technological capabilities (Brynjolfsson and Hitt, 2000). Moreover, new business processes, new skills, and new organizational and industry structures are major drivers of the contribution of information technology. Therefore, one of the major concerns among organizations is to explore how an investment in IT and its diffusion would affect their productivity, a topic that has been the subject of much debate in the researchers' community in recent decades.

The research community is divided on the issue of the benefits of IT, and several studies have stressed the need for theoretical models that trace the path from IT investments to business value. In light of this argument, the development of the process-oriented perspective throws some light on this aspect as it examines the effects of IT on intermediate business processes. This view has gained additional support from the theoretical developments in process innovation and business process engineering that are well documented within the academic literature. A number of studies (Clemons and Row, 1991; Mata et al., 1995) have reported that investment in IT can be easily duplicated by rival organizations; thus, just investing in IT never gives an organization a competitive advantage, but rather how firms leverage their investments to create unique IT resources and skills determines organizations' effectiveness and competitive capability.

3.3.2 IT in Process Management

It has been made clear from the arguments so far how IT has become a significant and central part of the organization's performance. Based on the research evidence presented earlier, we would suggest organizations should not just focus on IT investments, but rather continuously identify the ways in which IT can be developed as a unique resource since the ultimate aim of any organization is to outdo its rivals. Now let us focus on the understanding of the role of IT in managing the business processes. Business processes could be viewed as being comprised of two dimensions, operational processes and managerial processes, and IT does have implications for both operational and management processes. In a manufacturing setting, operational processes are affected by a number of different technologies, including robotics, computer-aided design (CAD), computer-aided manufacturing (CAM), flexible manufacturing, data capture and storage devices, imaging, and work flow systems. Here IT can play a significant role in improving the operational efficiency through automation, or it can enhance the effectiveness and reliability of operational processes by linking them together (Mooney et al., 1995). On the other hand, when it comes to improving management processes, IT can lead the way by improving the efficiency and effectiveness of communication through the availability and communication of information through e-mails, databases, and video conferencing. IT also acts as a helpful tool to integrate the different business units through end-to-end linking of value chains of one business unit with those of another business unit, thus supporting the interorganizational business processes. Particularly in the case of a related diversified company where plenty of value chain matchups exist among the different business units, IT can provide excellent support to business processes. IT can be of great assistance in overseeing operational and managerial processes if one can establish a synergy between information systems and business processes. The success or failure of any organization's use of IT does depend, however, on the managers' ability to understand and implement a process view.

IT has also become an essential and integral part of process reengineering efforts, primarily as an enabler of new operational and management processes, and thus improving the value-added work flow. IT allows organizations to perform business processes more proficiently, such as through automation, knowledge management, tracking, a reduction in intermediaries, and providing project management skills. Furthermore, when competently applied, IT can provide support for the intermediate processes, which, when taken together, comprise the execution of an organization's strategy. Additionally, IT can be used to integrate both hardware and software elements in an organization that aims to reduce the lead time at various places (Gunasekaran and Nath, 1997). Practitioners also need to understand that IT helps to improve the communication between various functional areas within the organization, leading to cooperative supported work for an improved productivity and quality. Thus, it is clearly evident that IT has a significant role to play in managing processes, and organizations need to develop unique IT capabilities in order to sustain their competitive advantage.

3.4 Identifying Key Processes

So far we have discussed and understood the notion of managing business processes and have seen how IT has emerged as an integral part of the organization's process improvement plan. Discussions presented earlier have also established the significance of having a better understanding of business processes, without which it is rather hard for the management to improve the organizational performance. Organizations are well aware of the fact that today's business environment is quite competitive, and meeting quality requirements has become a normal means of competition. This is why we are putting an emphasis on a well-established and implemented quality management system (QMS). Meeting quality requirements is a lot easier said than done however. This condition worsens for organizations if their core business processes are widely dispersed and inconsistent. In core business processes we refer to processes that are essential to the delivery of outputs and achieving business goals. The consistency of the core/key business processes is essential for organizations to respond quickly to the changing market conditions. Failure to respond quickly can lead to significant losses in market share and profitability, and in some cases organizations can even completely lose the competitive battle. Therefore, organizations need to distinguish their core business processes from the other processes.

So how do organizations identify their core/key business processes? We have argued that core business processes are central to the delivery of output and the organization's business objectives. Being central to delivery output, core processes will have a significant impact on the success of an organization, whereas being aligned to business objectives, core processes deliver results aligned to specific and measurable business goals. Thus, it is clearly evident that core business processes are real value-creating processes in the organization. From the process perspective organizations can be defined as a combination of transformational and transactional processes. An organization's transformational processes, meaning the conversion of inputs to outputs and transactional processes, namely, the exchange of outputs for inputs, can be separated into a number of commonly accepted business functions, such as the production, distribution, sales, billing and collection, accounts receivable, purchasing, accounts payable, product development, legal, personnel, and financial processes. But the challenge remains the same: How can an organization identify those key processes?

It is well known that all organizations are positioned somewhere between the suppliers and customers. Therefore, a business process cannot just be simply prioritized keeping in mind its closeness to the customer end, since for organizations the supplier relationship is as important as the customer relationship. The only way to counter this problem is to map existing processes to identify the outputs being delivered and then work backward from there to identify the processes that yield these outputs. The organization's critical success factors are normally assisted by the core processes, which act as drivers of key performance indicators. Mapping the business processes provides a clear link between an organization's processes and related outputs. The visualization of processes also helps an organization to identify the areas of importance that otherwise would remain ignored. It is also important to identify the cost associated with the different processes, as that can assist organizations in further identifying some key processes that may be financially very important. Therefore, charting/process mapping eases the task of the management in identifying and making decisions related to cost reductions, improving operations, or reinvesting in some different processes/functions.

Once the core processes are identified, it is an ideal practice to rank them in order of their importance in terms of achieving businesses objectives and output delivery. The organizations also need to identify business activities that support these core processes. The organizations' focus should then be on improving these core processes based on their priority. The improvement can be achieved by investigating and removing possible obstacles and educating employees on what the core business process is and how it will provide assistance to their respective areas. Once organizations start to follow such practices or adopt a culture of identifying core processes and improving them, they then continue to be critical success factors that give them a significant competitive advantage. So now we realize that identifying the core process is central to an organization's success and performance. From a quality improvement perspective as well, the identification and management of core processes is vital. Now let us focus on the significance of value stream mapping.

3.5 Value Stream Mapping and Modeling

Previous sections have discussed briefly the role of information technology in managing business processes and highlighted the significance of identifying the core business processes. From the discussions so far it is obvious that modern organizations are left with no alternative but to continuously seek opportunities to create and deliver value to the customer. However, when we argue about creating value, organizations can only create and deliver value when they understand which activities within their organization are particularly important in creating the value and what activities are not adding any particular value. The concept of value chain and value networks can come to the rescue of organizations and assist them in how they understand this notion of value creation. In plain words, the value chain is a combination of all the various activities that an organization performs internally in order to create value for customers. The organization's value chain can be classified into two broad categories: primary activities that are directly concerned with the creation and delivery of products or services and are principal in creating value, and secondary activities that are supporting activities to facilitate and enhance the performance of the primary activities. For example, for a manufacturing company primary activities would include activities related to inbound logistics, operations, outbound logistics, marketing and sales, and service activities. On the other hand, examples of the secondary or support activities would include procurement, technology development, human resource management, and infrastructure. The classification and separation of activities into primary and secondary activities not only assists organizations in understanding whether a set of activities provides any benefit to their final products or service offerings, but also helps them to understand their cost structures.

Driven by the concept of the value chain, a terminology that is well known among academics and business practitioners concentrating on performance improvement is value stream mapping (VSM). VSM was developed in 1995, with an underlying rationale of providing assistance to researchers and business practitioners to identify wastes in individual value streams and find an appropriate way to remove them. A value stream is a collection of value-added as well as non-value-added activities that are required to bring a product or a group of products through the main flows, starting with raw material and ending with the customer (Rother and Shook, 1999). The term *main flows* refers to the information and material flows that are across the whole value chain. The prime goal of VSM is twofold: first to identify all the different types of wastes (non-value-adding activities) that exist in the value stream, and then to take necessary actions to try to eliminate these wastes. Thus, by identifying the different value-added and non-value-added activities in the value stream, VSM aims to eliminate the wasteful activities and align the production with the demand.

Nonetheless, in the manufacturing field there are certain activities that are non-value-adding in nature but necessary, such as unpacking/unloading deliveries and transferring a tool from one hand to another. These necessary but non-value-added activities can be eliminated, but doing so will require extensive changes in the operating systems, and sometimes it is not feasible to make those changes immediately. The goal of eliminating wastes originates from Lean manufacturing principles, and the choice of wastes in manufacturing operations originates from the Toyota Production System (TPS) developed in the 1980s. TPS defines seven commonly accepted wastes, also referred to as *muda*: overproduction, waiting time, transport costs, unnecessary or complicated processing, excess inventory, unnecessary motion, and defects. VSM consists of five phases:

1. Selection of a product family
2. Current state mapping (CSM)
3. Future state mapping (FSM)
4. Defining a working plan
5. Achieving the working plan

Therefore, in VSM, in order to identify the value-adding and non-value-adding activities in the value stream, the first step is to choose a particular product or product family as the target for improvement. The second step is to draw a current state map of each value stream of a specific product or product family within a plant. From the business practitioner's viewpoint this step involves an understanding of how processes are being carried out currently. Also at this stage, it is important to identify and analyze the seven sources of wastes. The third step is to create a future state map, i.e., to have a view of how the system would look after the inefficiencies have been removed. This is done by answering a set of questions on issues related to efficiency, and on technical implementation related to the use of Lean tools. The mapping of the value stream activities from raw materials to end consumer helps organizations to evaluate the overall efficiency of the entire value stream by determining performance indicators such as total lead time, total value-adding time, number of inventory turns, level of defects at each stage, occurrences of the bullwhip effect, and total miles traveled. Based on the first three steps, the next step involves creating a work plan with the aim of eliminating any non-value-added activity, and the fifth and final step involves executing the work plan and achieving goals.

As evident from the discussion presented in earlier sections, the understanding and improvement of processes is essential to the efficiency of VSM. There are seven initial tools used for VSM derived from a variety of functional and academic backgrounds, such as engineering, action research, system dynamics, and operations management. These seven tools are process activity mapping, supply chain response matrix, production variety tunnel, quality filter mapping, demand amplification mapping, decision point analysis, and physical structure mapping. Research evidences have shown that VSM is a suitable tool for redesigning production systems. In general, a complementary tool is needed along with VSM that can quantify the gains during the early planning and assessment stages. An obvious tool is simulation, which is capable of generating resource requirements and performance statistics while remaining flexible in relation to specific organizational details. There are also other tools of process improvement, such as process mapping and the Icam DEFinition Zero (IDEF0) method. In summary, VSM could be a very useful tool to improve processes. The next section elaborates the process improvement agenda.

3.6 Process Improvement Agenda

Discussions so far presented in this chapter are urging organizations to develop a process improvement agenda/plan as a priority. With process improvement we mean to say that organizations need to seek ways to make things better on a continual basis, not just responding to the ongoing problems and crises. In most organizations, whenever a problem arises the "blame game" starts, leading to criticism of either workers or managers, or even situations where people are fired from their jobs. But the notion of process improvement is about setting aside the customary practice of blaming people for problems or failures, and instead identifying ways to resolve the problem and continuously looking to improve working practices. Thus, organizations willing to improve their performance through optimizing their underlying processes need to devise a process improvement agenda.

Generally, organizations looking to improve their working practices, or processes often take a problem-solving route where they simply attempt to fix what has been broken. The disadvantage of this approach, however, is that often organizations fail to track the root cause of the problem. The failure to address the root cause of the problem leads to the repetition of similar problems in the future. Leaders play a crucial role in this process of the elimination of wastes, as they are the ones who can build a culture within the organization where each employee attempts to examine the source of a problem rather than just fix it and move ahead. Therefore, organizations following a process improvement agenda encourage their employees to analyze the source of the problem by understanding all of the conditions that can potentially lead to such situations. Consequently, the move toward process improvement is about the collective teamwork that eliminates wasteful activities and streamlines productivity.

Leaders in organizations are responsible for driving the process improvement initiatives across all the levels, i.e., from top to bottom. Leaders should make sure their employees receive the required training that will enable them to carry out their process improvement efforts efficiently and effectively. But instilling a new culture within an organization is very challenging, and often leaders struggle to encourage employees to think beyond the accustomed way of doing things. The *Handbook for Basic Process Improvement* (1996) suggests 14 steps to improve the processes:

Step 1: Select a process and establish the process improvement objective.
Step 2: Organize the right team.
Step 3: Flowchart the current process.
Step 4: Simplify the process and make changes.
Step 5: Develop a data collection plan and collect baseline data.
Step 6: Is the process stable?
Step 7: Is the process capable?
Step 8: Identify the root causes of lack of capability.
Step 9: Plan to implement the process change.
Step 10: Modify the data collection plan, if necessary.
Step 11: Test the change and collect data.
Step 12: Is the modified process stable?
Step 13: Did the process improve?
Step 14: Standardize the process and reduce the frequency of data collection.

These 14 steps of the process improvement model enhance the team's process knowledge, broaden their decision-making options, and increase the likelihood of satisfactory long-term results. Many of the steps in the business process improvement model (steps 8 to 14) are part of the plan, do, check, act (PDCA) cycle, frequently used by organizations that are following quality improvement initiatives. Therefore, we would like to emphasize that organizations should first instill a process improvement-oriented working culture and train their employees very well if they really want to be successful. They also need to have a well-defined and well-documented process improvement agenda. Success stories suggest that even though an organization has a good agenda, if it fails to execute it very well, then the outcome can be problematic. On the other hand, even if an organization has a poor agenda but it has been very well executed, then the results can be acceptable. Thus, organizations need to give the same amount of attention to designing a process improvement agenda as they do to its execution.

The discussions presented earlier on the subject of VSM also suggest that the focus of any process improvement initiative should be on eliminating wastes, whether overproduction, waiting time, processing time, or defects. Organizations are normally aware of these process improvement goals, however, so the next question that comes to mind is: Who is going to be a winner? The organization that will emerge as a winner will be the one that can achieve these process improvement goals more cheaply, quickly, easily, and safely. Therefore, an organization's process improvement agenda needs to be well defined and well assessed (from cost, safety, convenience, and feasibility perspectives) in a manner that merits good execution.

3.7 Summary

This chapter has elaborated on the need for, and advantages of, managing processes efficiently, and along the way it has highlighted the benefits of developing IT competence, identifying core processes, and value stream mapping in the overall performance improvement of an organization. To clarify the need for managing processes, we have first put an emphasis on why this needs to be done. We have tried to explain to organizations how improving their processes can lead to an improvement in overall performance. To clarify further, we have taken an approach of first defining the concept of process management. Thereafter, we have highlighted the importance of process management. We have identified that efficient and effective process management is central to quality improvement initiatives. The chapter also detailed the significance of IT in managing processes and stressed the need for developing IT competence. We have argued that core business processes are real value-creating processes, and organizations need to identify and improve them to strengthen their competitive position. A brief discussion on value stream mapping (VSM) has also been provided. We have concluded this chapter by suggesting that the development of a well-defined process improvement agenda must be a priority for organizations. A

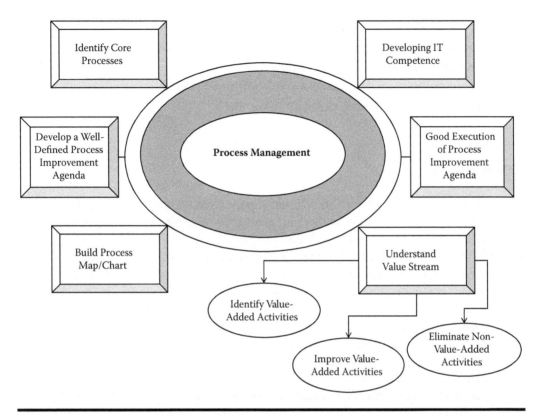

Figure 3.1   Chapter summary illustration.

summarized view of this chapter is presented in Figure 3.1. In Chapter 4 we highlight the importance and necessity of integrating the QMS and business processes diagnostic into the organization's business plan and strategy.

3.7.1 Key Points to Remember

■ Organizations need to establish a well-designed business process that can achieve internal efficiency and external effectiveness.
■ To understand and visualize the processes in a better way, organizations need to map the processes.
■ The essence of managing processes is to identify the means of performing tasks in meaningful, efficient, and effective ways.
■ To build IT competence, organizations need to be familiar with their capacities and well aware of the limitations and impact of computer technology.
■ Identifying the core process is central to the organization's success and performance.

■ The objective of VSM is to identify all the different types of waste (non-value-adding activities) that exist in the value stream, and then take the necessary actions to try to eliminate these wastes.

■ Organizations need to have a well-defined process improvement agenda, and it must be carefully executed.

References

Bharadwaj, A. S. (2000). A resource-based perspective on information technology capability and firm performance: an empirical investigation. *MIS Quarterly*, Vol. 24, No. 1, pp. 169–196.

Brynjolfsson, E., and Hitt, L. M. (2000). Beyond computation: Information technology, organizational transformation and business performance. *Journal of Economic Perspectives*, Vol. 14, No. 4, pp. 23–48.

Chan, S. L. (2000). Information technology in business processes. *Business Process Management Journal*, Vol. 6, No. 3, pp. 224–237.

Clemons, E. K., and Row, M. C. (1991). Sustaining IT advantage: The role of structural differences. *MIS Quarterly*, Vol. 15, No. 3, pp. 275–294.

Gunasekaran, A., and Nath, B. (1997). The role of information technology in business process reengineering. *International Journal of Production Economics*, Vol. 50, No. 2/3, pp. 91–104.

Konar, E., Kraut, A., and Wong, W. (1986). Computer literacy: With ask you shall receive. *Personnel Journal*, Vol. 65, No. 7, pp. 83–86.

Kumar, V., Smart, P. A., Maddern, H., and Maull, R. S. (2008). Alternative perspectives on service quality and customer satisfaction: The role of BPM. *International Journal of Service Industry Management*, Vol. 19, No. 2, pp. 176–187.

Laguna, M., and Marklund, J. (2004). *Business process modelling, simulation, and design*. Pearson Education, Upper Saddle River, NJ.

Mata, F. J., Fuerst, W. L., and Barney, J. B. (1995). Information technology and sustained competitive advantage: A resource-based analysis. *MIS Quarterly*, Vol. 19, No. 4, pp. 487–505.

Melan, E. H. (1993). *Process management: Methods for improving products and services*. McGraw Hill, New York.

Mooney, J., Gurbaxani, V., and Kraemer, K. (1995). A process oriented framework for assessing the business value of information technology. At International Conference on Information Systems (ICIS), 1995 Proceedings, paper 3.

Rother, M., and Shook, J. (1999). *Learning to see: Value stream mapping to add value and eliminate muda*. Lean Enterprise Institute, Brookline, MA.

The Handbook of Basic Process Improvement. (1996). CINCPACFLTINST 5224.2.

Further Suggested Readings

Milgrom, P., and Roberts, J. (1990). The economics of modern manufacturing: Technology, strategy, and organization. *American Economic Review*, Vol. 80, No. 3, pp. 511–528.

Nickols, F. (1998). The difficult process of identifying processes. *Knowledge and Process Management*, Vol. 5, No. 1, pp. 14–19.

Rockart, J. F., Earl, M. J., and Ross, J. W. (1996). Eight imperatives for the new IT organization. *Sloan Management Review*, Fall 1996, pp. 43–55.

Chapter 4

Quality Management Systems and Business Processes Diagnostic

4.1 Introduction

Understanding the current situation of an organization's quality management system (QMS) and business processes is important since it can prove instrumental in determining the quality of subsequent management decisions to effectively design or improve a QMS. In this chapter we propose a methodology that provides overall guidelines to help organizations carry out a diagnosis of the status of their QMS and business processes. The methodology is based on the definition and understanding of the maturity level of a company's QMS and on the assessment and identification of its strengths and opportunities for improvement in its core business processes. The methodology also integrates quality audits as a means to providing further information about the QMS and its compliance with the standards of customers, suppliers, partners, collaborators, the industry sector, or even government. We conclude this chapter by highlighting the importance and necessity of integrating the QMS and business processes diagnostic into the organization's business plan and strategy to create an improvement agenda and key suggestions for its deployment.

4.2 Defining the QMS Maturity Level

The diagnosis of a QMS and business processes must start by defining and understanding the maturity of the organization's structure, procedures, processes, and resources dedicated to ensure that their products and services satisfy their customers' expectations. In this text, we refer to maturity as the degree of knowledge, use, effective deployment, and concrete positive results obtained from a company's QMS. Dale and Lascelles' (1997) six-level categorization model provides a simple tool for evaluating and understanding the current organizational situation in reference to the degree of maturity of its QMS. This model identifies six levels in the adoption of Total Quality Management (TQM) principles, which can be used as a platform for performing the assessment. Based on this model, the six levels of categories an organization may fall under are (1) uncommitted, (2) drifters, (3) tool pushers, (4) improvers, (5) award winners, and (6) world-class (see Figure 4.1).

Table 4.1 presents a maturity diagnostic instrument (MDI), which we have adapted and designed based on Dale and Lascelles' (1997) model. In addition to helping measure the maturity of an organization's QMS, this instrument can also help set a general before and after improvement comparative base and identify specific limitations and thus business improvement needs. As this model has been combined with a Likert scale in this instrument, it can also procure a level of development measure for every specific subcategory.

When using the instrument, only one number (e.g., 1, strongly agree; 2, agree; 3, agree slightly; etc.) has to be circled for each of the 84 subcategories in Table 4.1. This will indicate the assessment team's perception regarding the position of the company in relation to each of these subcategories. Once this has been done, the numbers that have been circled have to be

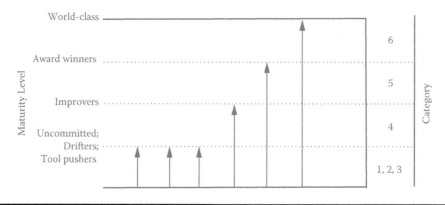

Figure 4.1 Illustration of the six-level categorization model of Dale and Lascelles (1997).

Table 4.1   Maturity Diagnostic Instrument (MDI)

Subcategory	Strongly Agree	Agree	Slightly Agree	Neutral	Slightly Disagree	Disagree	Strongly Disagree
1. Quality improvement (QI) initiatives *are not* only carried out to achieve ISO 9000 registration or comply with customer requirements.	1	2	3	4	5	6	7
2. Initial enthusiasm after implementing a quality management system (QMS) or QI program *does not* fade over time.	1	2	3	4	5	6	7
3. Organization holds an ISO 9000 certification (or is close to obtaining it).	1	2	3	4	5	6	7
4. Organization recognizes that the effective implementation of a QMS requires cultural change.	1	2	3	4	5	6	7
5. Organization has a culture where quality *is not* dependent on the commitment and drive of a limited number of individuals.	1	2	3	4	5	6	7
6. A total integration of continuous improvement (CI) and business strategy to delight customers exists.	1	2	3	4	5	6	7
7. Organization *does not* only apply quality management (QM) tools and techniques due to customers' presence, monitoring, and pressure.	1	2	3	4	5	6	7
8. Organization *has not* expressed disappointment about the current QMS.	1	2	3	4	5	6	7
9. Organization employs a selection of quality management tools (e.g., statistical process control (SPC), quality circle (QC), failure mode and effects analysis (FMEA), mistake proofing, quality improvement groups).	1	2	3	4	5	6	7
10. Organization recognizes the importance of customer-focused CI.	1	2	3	4	5	6	7
11. All employees are involved in CI.	1	2	3	4	5	6	7
12. Organization's purpose and values are defined and communicated at all levels.	1	2	3	4	5	6	7

Table 4.1 *(Continued)* **Maturity Diagnostic Instrument (MDI)**

Subcategory	Strongly Agree	Agree	Slightly Agree	Neutral	Slightly Disagree	Disagree	Strongly Disagree
13. Not only does the quality department drive the QMS and maintain ISO certification, but all staff participate and have concern for quality.	1	2	3	4	5	6	7
14. Organization *is not* susceptible to the adoption of the latest QM fads.	1	2	3	4	5	6	7
15. Organization *does not* tend to look for the latest QI approaches/tools for a "quick fix."	1	2	3	4	5	6	7
16. Senior management shows commitment toward QI through both leadership and personal actions.	1	2	3	4	5	6	7
17. A number of successful organizational changes have been made.	1	2	3	4	5	6	7
18. Organization has developed and applied a unique success model.	1	2	3	4	5	6	7
19. Success of quality initiatives *is not* linked to the success of external audits only.	1	2	3	4	5	6	7
20. Management teams *do not* try a variety of approaches in response to the latest quality management (QM) fads.	1	2	3	4	5	6	7
21. All senior management members are committed to the organization's QMS.	1	2	3	4	5	6	7
22. Organization has formulated a quality strategy and implemented, at least, a good portion of it.	1	2	3	4	5	6	7
23. Business procedures and processes are efficient and responsive to customer needs.	1	2	3	4	5	6	7
24. Organization places a positive value on internal and external relationships (e.g., with employees, customers).	1	2	3	4	5	6	7
25. QM *is not* considered a contractual requirement and an added cost.	1	2	3	4	5	6	7

Table 4.1 *(Continued)* **Maturity Diagnostic Instrument (MDI)**

Subcategory	Strongly Agree	Agree	Slightly Agree	Neutral	Slightly Disagree	Disagree	Strongly Disagree
26. Senior management *does not* assume that CI occurs naturally or is self-sustained.	1	2	3	4	5	6	7
27. CI efforts are not only concentrated in manufacturing/operations departments, but also in other departments of the organization.	1	2	3	4	5	6	7
28. A problem-solving infrastructure and a proactive QMS are in place.	1	2	3	4	5	6	7
29. Process improvement results are measurable and carried out through effective cross-functional management.	1	2	3	4	5	6	7
30. Organization works in partnership with stakeholders.	1	2	3	4	5	6	7
31. Priority is given to QI in terms of time and allocation of resources.	1	2	3	4	5	6	7
32. Organization has adopted different quality philosophies (e.g., Deming, Crosby, Juran, SPC, International Organization for Standardization (ISO), TQM, Six Sigma).	1	2	3	4	5	6	7
33. A QMS exists and the data it provides are used to their full potential.	1	2	3	4	5	6	7
34. A long-term and company-wide education/training program is in place.	1	2	3	4	5	6	7
35. Strategic benchmarking is practiced at all levels.	1	2	3	4	5	6	7
36. QMS helps to identify opportunities to improve the ability of the company to satisfy its customers.	1	2	3	4	5	6	7
37. Corrective actions *are not* only taken in response to customer complaints.	1	2	3	4	5	6	7
38. Continuous improvement is perceived as a strategy, not as a program only.	1	2	3	4	5	6	7

Table 4.1 *(Continued)* **Maturity Diagnostic Instrument (MDI)**

Subcategory	Strongly Agree	Agree	Slightly Agree	Neutral	Slightly Disagree	Disagree	Strongly Disagree
39. Long-term results in all organizational aspects (as opposed to short-term results regarding product output and quality only) are expected.	1	2	3	4	5	6	7
40. Individual staff carry out improvement activities within their own spheres of influence and on their own initiative.	1	2	3	4	5	6	7
41. A system for internal and external performance measurement is in place.	1	2	3	4	5	6	7
42. Organization is constantly looking to identify new/ more products, services, or characteristics that will increase customer satisfaction.	1	2	3	4	5	6	7
43. Support to solve problems *is not* based on their impact on sales/turnover only.	1	2	3	4	5	6	7
44. A plan for effectively deploying a QMS exists.	1	2	3	4	5	6	7
45. Processes *do not* have considerable potential for improvement.	1	2	3	4	5	6	7
46. Importance of staff involvement in CI is recognized, communicated, and celebrated.	1	2	3	4	5	6	7
47. Employees at all levels reflect a participate culture.	1	2	3	4	5	6	7
48. A QI culture is no longer dependent on top-down drives, but it is also driven laterally through the whole organization.	1	2	3	4	5	6	7
49. Quality of design has a high priority.	1	2	3	4	5	6	7
50. Management *is not* oversusceptible to outside intervention and *does not* easily get distracted by the latest QM and CI fads.	1	2	3	4	5	6	7
51. All parts of the organization believe that the current QMS is effective.	1	2	3	4	5	6	7

Table 4.1 *(Continued)* **Maturity Diagnostic Instrument (MDI)**

Subcategory	Strongly Agree	Agree	Slightly Agree	Neutral	Slightly Disagree	Disagree	Strongly Disagree
52. Benchmarking studies have been initiated and the results used for CI.	1	2	3	4	5	6	7
53. Management practices a culture of empowerment.	1	2	3	4	5	6	7
54. The vision of the entire organization is aligned to the voice of the customer.	1	2	3	4	5	6	7
55. Organization has made an acceptable investment in quality education and training.	1	2	3	4	5	6	7
56. Quality department has a high status within the organization.	1	2	3	4	5	6	7
57. Momentum of improvement initiatives is easy to sustain.	1	2	3	4	5	6	7
58. Organization has QI champions among some senior management members.	1	2	3	4	5	6	7
59. Current QMS is sincerely viewed by all employees as a way of managing the business to satisfy and delight customers, both internal and external.	1	2	3	4	5	6	7
60. Total quality is the organization's "way of life" and "way of doing business."	1	2	3	4	5	6	7
61. Senior management takes responsibility for CI/QI activities.	1	2	3	4	5	6	7
62. The "born and died" of improvement teams *is not* a constant phenomenon.	1	2	3	4	5	6	7
63. Training on quality tools is aimed at persons who can influence their further application.	1	2	3	4	5	6	7
64. Trust between all levels of the organization exists.	1	2	3	4	5	6	7
65. Perception of stakeholders of the company's performance is surveyed and acted on to drive improvement actions.	1	2	3	4	5	6	7

Table 4.1 *(Continued)* **Maturity Diagnostic Instrument (MDI)**

Subcategory	Strongly Agree	Agree	Slightly Agree	Neutral	Slightly Disagree	Disagree	Strongly Disagree
66. Quality values are fully understood and shared by employees, customers, and suppliers.	1	2	3	4	5	6	7
67. Organization has had positive previous experience with ISO, TQM, or other quality management approaches.	1	2	3	4	5	6	7
68. Cultural changes have taken place after the implementation of CI/QI programs.	1	2	3	4	5	6	7
69. Quality tools and techniques are implemented strategically and not only reactively and when necessary.	1	2	3	4	5	6	7
70. There is low preoccupation with numbers (e.g., financial measures).	1	2	3	4	5	6	7
71. Results of improvement projects are effectively utilized.	1	2	3	4	5	6	7
72. Each person in the organization is committed, in an almost natural way, to seek opportunities for improvement.	1	2	3	4	5	6	7
73. There *is not* an overwhelming emphasis on the achievement of financial measures.	1	2	3	4	5	6	7
74. Appropriate knowledge of the current QMS exists.	1	2	3	4	5	6	7
75. Meeting output targets *is not* the only key priority for the majority of managers; there are no conflicts between the production/operations department and the quality department.	1	2	3	4	5	6	7
76. QI drives and direction *do not* rely only on a small number of individuals.	1	2	3	4	5	6	7
77. All things are done right the first time.	1	2	3	4	5	6	7
78. Dependability is emphasized throughout the organization.	1	2	3	4	5	6	7

Table 4.1 *(Continued)* **Maturity Diagnostic Instrument (MDI)**

Subcategory	Strongly Agree	Agree	Slightly Agree	Neutral	Slightly Disagree	Disagree	Strongly Disagree
79. There is a long-term plan for corrective actions for reoccurrence of problems.	1	2	3	4	5	6	7
80. Self-assessment is performed and improvements identified are addressed.	1	2	3	4	5	6	7
81. The organization has a flexible QMS not only designed to fulfill customer regulations.	1	2	3	4	5	6	7
82. If key directors/managers/individuals leave, business mergers occur, organizational restructuring takes place, etc., there *is no* danger of losing momentum or failure in terms of QM/QI initiatives.	1	2	3	4	5	6	7
83. QMS is effective and it does help to identify opportunities to improve the ability of the company to satisfy its customers.	1	2	3	4	5	6	7
84. Waste is not tolerated.	1	2	3	4	5	6	7

transferred to the corresponding columns of the scoring table (Table 4.2). Subsequently, they need to be added, and the result of each sum divided by 14. This will give comparable scores, where the highest score will indicate the organization's status of quality maturity and category (e.g., "uncommitted," "drifters," etc.) in reference to the assessment model.

4.2.1 Interpretation and Diagnosis

An important consideration is the diagnosis made based upon the data interpretation. Dale and Lascelles (1997) recognize that some organizations may fall midway between some of the categories, while others may display hybrid quality structures, procedures, processes, and resources found in two or more groups. Defining a specific category based on the highest score will provide a general overview of the QMS status. However, a simple but more meaningful diagnosis would be to assess the amount of variance for each of the 84 subcategories in relation to a score of 4, which is the neutral point.

Table 4.2 Scoring Table

1. Uncommitted		2. Drifters		3. Tool Pushers		4. Improvers		5. Award Winners		6. World-Class	
Question	Value	Question	Value	Question	Value	Question	Value	Question	Value	Question	Value
1		2		3		4		5		6	
7		8		9		10		11		12	
13		14		15		16		17		18	
19		20		21		22		23		24	
25		26		27		28		29		30	
31		32		33		34		35		36	
37		38		39		40		41		42	
43		44		45		46		47		48	
49		50		51		52		53		54	
55		56		57		58		59		60	
61		62		63		64		65		66	
67		68		69		70		71		72	
73		74		75		76		77		78	
79		80		81		82		83		84	
Total		Total		Total		Total		Total		Total	
÷14		÷14		÷14		÷14		÷14		÷14	
Average		Average		Average		Average		Average		Average	

Scores above 4 would indicate a problem with a specific quality process or practice. The closer the score is to 7, the more severe the problem would be. Scores below 4 indicate the lack of a problem, with a score of 1 indicating an optimum quality process or practice. Although the MDI proposed provides a simple mechanism by which to evaluate and define the current status of an organization's QMS at a specific point in time, the real potential of this instrument is that it can serve as a measure of improvement. For example, several assessments can be carried out at different points in time to compare the scores in each category and subcategory; if the score increases, this would indicate that the organization has made some progress in that particular subcategory or moved within the six-level scale of Dale and Lascelles (1997).

4.2.2 Performing the Assessment Using the MDI

The evaluation of QMS maturity using the MDI should be carried out by a multidisciplinary team comprised of staff from different functional areas (e.g., quality, production, materials, human resources) and different levels (e.g., top and middle management, supervisors, shop floor operators) of the organization. This will ensure a thoughtful and hence reliable assessment wherein different perspectives and feelings are taken into consideration. The evaluating team should also have sufficient credibility to ensure that the organization "buys in to" the QMS maturity assessment and its results. On the other hand, to reduce subjectivity and avoid an inaccurate interpretation of the results, it is recommended that the same team perform the evaluation of the maturity of the organization's QMS. Although this will not completely eliminate the subjectivity of the MDI, it will help reduce variability in the assessors' perceptions, and thus improve the reliability of the quality maturity assessment.

4.3 Identifying Strengths and Opportunities for Improvement in the Organization's Business Processes: A Self-Assessment Approach

Once the maturity of a company's QMS has been defined, the next stage in diagnosing the status of its QMS and business processes is to determine the organization's strengths and opportunities for improvement in its core business processes. By this stage, the MDI presented in the previous section would have already provided the organization with some insight on its strengths and opportunities for improvement. However, a more thorough measure and analysis involving different aspects of the organization's business activities and core processes are required to achieve this. A self-assessment approach based on the use of a business excellence model (BEM) can provide an organization with a powerful approach to achieving this. The use of the BEMs, as previously reviewed in Chapter 2, has quickly moved from one of mere award participation to a more holistic approach employed by organizations to self-assess their operations. In general terms, self-assessment provides organizations with a detailed picture of their business processes and helps identify areas in need of improvement. Although this can be considered the main objective of a self-assessment process and a prime element for selecting, designing, implementing, and improving a QMS, there are some other benefits associated with the use of BEMs when employed as a self-assessment method. Some of the most important and common benefits are summarized in Table 4.3.

4.3.1 A Best-Practice Approach for Conducting a Self-Assessment Process

Some authors and experts propose several approaches to effectively carrying out a self-assessment exercise. Figure 4.2 presents a comparison of some of these approaches.

Based on these methods, the literature and practical experience, we propose the following approach for conducting the self-assessment process:

4.3.1.1 Stage 1: Setting the Organizational Environment for the Self-Assessment Process

Preparing the organization to positively respond and contribute to the self-assessment process is essential to its success. For this reason, a contributive environment must be established by performing some preparatory work before conducting the self-assessment process. This preparatory work should include

■ The formation of a review committee comprised of top management employees able to directly communicate with the company's CEO, influence strategic decisions, carry out follow-up actions, and correct the direction if necessary (Antony and Preece, 2002). It is also important for

Table 4.3 Some Important and Common Benefits of Self-Assessment

Benefit	Category	Source
Improves operational and financial performance	Business results	European Foundation for Quality Management; European Center for Quality Management (Porter and Tanner, 1998; Gadd, 1995)
Improves customer satisfaction		
Links business results with what organizations have to do to achieve such results		
Award-winning potential, which enhances organization's image and reputation		
Increases awareness of quality through the organization	Culture	
Improves focus and involvement of senior management and staff in CI		
Allows managers a broader understanding of the business		
Promotes strategic action planning	Process management	
Provides a structured and rigorous approach to improve business operations		
Provides consistency in the direction of the organization and consensus on what needs to be done		
Encourages integration of quality-oriented initiatives		
Enforces a process management perspective and links processes to results		
Provides an assessment based on facts and not opinions	Benchmarking	
Helps to more effectively measure the progress of an organization		
Helps to prioritize improvements		
Enables a comparison between departments and divisions and against other organizations		

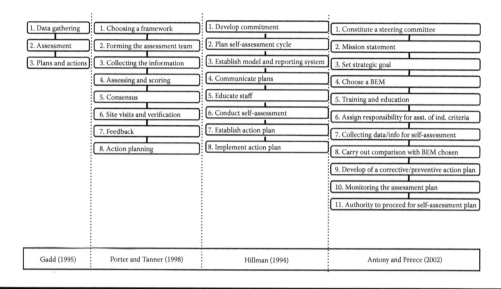

| Gadd (1995) | Porter and Tanner (1998) | Hillman (1994) | Antony and Preece (2002) |

Figure 4.2   Some approaches to self-assessment found in the literature.

this committee to act not only as a reviewer but also as a champion of the self-assessment process by creating a sense of urgency and demonstrating a need for the process to take place.

■ Gaining commitment from all the organization's employees to ensure that the self-assessment process is not perceived to be yet another audit (Hillman, 1994). In a self-assessment process the organization's performance and improvement are evaluated against a model for continuous improvement (CI). By contrast, in traditional audits checks are carried out to assess whether the organization complies with certain procedures laid out in manuals or standards.

■ A review of the organization's mission statement, or creation of one, to make sure that it is based on important values in regard to its customers (e.g., quality, flexibility, agility, dependability), and that it appeals to the company's stakeholders (Antony and Preece, 2002).

Some other factors include the following:

■ Ensuring commitment and involvement of top management, and relevant functional areas, in the design and development of the self-assessment instrument

■ Ensuring commitment from top management to dedicate the needed resources (e.g., time, personnel, finances, information, consultants) during the self-assessment process

■ Setting a communication channel through which to disseminate targets, execution progress, and results of the self-assessment process to all company employees

4.3.1.2 Stage 2: Selecting a BEM

Part of the responsibilities of the review committee would be to select the BEM that is most appropriate to carrying out the self-assessment process (some of the available and most common BEMs have been compared and reviewed in Chapter 2). As mentioned in this chapter, BEMs have different structures, focuses, and characteristics. For this reason, the selection of the BEM will depend upon the specific organization's characteristics and factors, such as size, industry, product/service, culture, quality maturity, geographical location, nationality, and experience with self-assessment. Porter and Tanner (1998) comment that "there is no 'best' framework, only an appropriate framework." Organizations may tend to adopt the most widely used or known BEM (e.g., Deming, Malcolm Baldrige, EFQM) or those available in their own countries. For example, a Mexican firm may be encouraged to adopt the Mexican Quality Model for Competitiveness (see Rocha-Lona et al., 2010). However, if main BEMs are thought not to be appropriate enough to assist the organization in the attainment of its strategic goals, a hybrid and more specific model, based on the criteria of the established models, can be created. Although a hybrid BEM would certainly serve the specific needs and strategic goals of an organization, it will not facilitate benchmarking with other organizations or benefit from an annual review and refinement of established models.

4.3.1.3 Stage 3: Forming and Training the Assessment Team

A wide range of areas that include leadership, people management, people satisfaction results, business analysis, and process management are addressed in a BEM's criteria. Realistically, no single person is likely to have an in-depth knowledge of all these areas. As a consequence, Porter and Tanner (1998) comment that it is a usual and suggested practice for the assessment team to be comprised of approximately six members from different functional areas of the organization. The assessment team in charge of performing the self-assessment process may be or may not be the same team in charge of evaluating the maturity of the organization's QMS using the MDI previously introduced. However, as the definition of the organization's maturity level and the self-assessment process are part of the methodology for diagnosing the status of the QMS and business processes, it would be preferred for the same team to perform both assessments. This will ensure some consistency and reduce the natural subjectivity involved in performing both evaluations.

Within the self-assessment team, a senior employee must assume the role of leader, whose main responsibility will lie in managing, motivating, and supervising the assessment team as well as acting as a direct link to the review committee. All personnel involved in the assessment team must be trained so as to ensure that they acquire the knowledge, expertise, and skills required to perform a systematic, reliable, consistent, and honest self-assessment. The knowledge, expertise, and skills should include

- A good degree of understanding of the BEM selected (e.g., its criteria and subcriteria, tools) and the strategic role of the assessment
- A good understanding of the overall self-assessment process and a deep understanding of the key steps or aspects most relevant to every team member
- An understanding of the cost and benefits of the self-assessment process and its role in the driving of CI
- A development of the team members' personal and technical skills and abilities to ensure a consistent assessment
- A development of the skills necessary to collect and analyze data as well as identify the gaps between the BEM's criteria and the current state of the organization
- A development of the skills necessary to write and provide clear and comprehensive feedback as well as to propose and implement the appropriate measures for bridging the gaps identified
- A clear understanding of the consequences associated with failure to take action

It is the responsibility of the assessment team to assess the organization's performance against each BEM criterion without introducing preconceived notions that may bias the self-assessment exercise. This can be a likely phenomenon when the assessment team is comprised of internal members, but is less common when the assessors are external to the organization. If the assessor is external, the organization will not need to form and train an assessment team. However, the review committee will need to ensure that the external assessor or assessment team has access to all the information and resources needed to do a proper and exhaustive assessment.

4.3.1.4 Stage 4: Collecting the Data and Information Needed for the Self-Assessment Process

In this stage of the self-assessment process, the assessment team is required to collect and present all the information needed to perform the organization's self-assessment against the selected BEM criteria and subcriteria. In terms of the data collection, this can be obtained through formal and informal interviews with staff, managers, and directors; questionnaires; examination of the company's documents; and information and perception of the assessment team members. Most of these data collection methods will require site visits, which will provide greater objectivity and a means of clarifying and verifying the data collected.

On the other hand, based on the Gadd's (1995) empirical research, an assessment team can capture and present the information using one of the following methods:

1. *Award-type position statement.* When an organization participates for a quality award such as the European Quality Award (EQA), it has to produce a document of no more than 75 pages in length that explains what the organization does and what it achieves. Gadd (1995) comments that while the preparation of this document is lengthy and time-consuming, some organizations still decide to produce it for self-assessment purposes, even if they do not intend to apply for the award. The empirical research carried out by Gadd (1995) suggests that the ways in which the data are collected to produce such a document vary considerably. For example, in some cases only one middle-level employee was in charge of the data collection, while in others only one director, or a group of directors, was in charge of such collection of data. Since a multidisciplinary assessment team should have already been formed and trained by this stage, the collection of the data needed to produce the document should be part of its responsibilities. This would make the data collection process more efficient and meaningful.

   Porter and Tanner (1998) suggest breaking down each BEM subcriterion or area into a set of questions and statements. For example, assuming that the organization has decided to use the EQA model, the assessment team can translate its criteria into questions such as (1) What does the organization currently do in this area? (2) How does it do it? (3) How widely used are these practices? (4) How is the organization's approach reviewed and what improvements are undertaken following a review? (5) How is the organization's approach integrated into normal business operations?

2. *Pro formas and worksheets.* An alternative to the preparation of submission documents is to capture and present the data in *pro formas* or worksheets. Gadd (1995) recognizes that although this method is much less exhaustive than the preparation of submission documents, it can still serve as an effective and less time-consuming alterative. In this case, responses to, for example, the questions previously stated can be recorded in the form.

3. *Discussion groups.* A third alternative that does not involve the previous collection of data or preparation of any documentation is the use of discussion groups. In this approach, the assessment team, based on their experience and perception of the organization, would be required to provide the information at the same meeting and time that the assessment takes place. This method would obviously require less preparation time and effort but does call for an in-depth knowledge of the organization's core business processes on the part of the assessment team, which would enable them to clearly and concisely describe these processes during the assessment meeting.

4.3.1.5 Stage 5: Assessing and Scoring

In this stage, every member of the assessment team must individually evaluate every criterion and subcriterion of the BEM selected and submit a score based on their perception of such criteria being implemented and practiced within the organization. Although scoring is a subjective exercise within the self-assessment process, the training previously provided to the assessment team members in stage 3 should contribute to the reduction of a natural variation of scoring. Main BEMs such as the EFQM and Malcolm Baldrige provide their own methods, guidelines, and charts for performing the scoring. It is therefore suggested that the scoring methods and tools proposed by the BEM selected in stage 2 be used. Alternatively, an organization may wish to simplify or adapt the scoring system of a main BEM to its own specific and direct needs and capabilities. The disadvantage of developing an in-house method for, in this case, scoring, is that (as previously discussed) it is more difficult to benchmark with other organizations that use a different scoring approach.

4.3.1.6 Stage 6: Achieving Consensus

The next stage in the self-assessment process is to reach a scoring consensus for each criterion and subcriterion evaluated as well as for the strengths and opportunities for improvement of the organization. This is because every member of the assessment team individually scores the organization's performance against the BEM criteria and subcriteria. Consensus is traditionally sought in a consensus meeting led by the assessment team leader. As a rule of thumb, and in order to conduct the consensus stage more efficiently, the EQA assessment indicates that if there is a less than 30% variation in the assessors' scores, then all the scores are simply averaged. This will provide an overall score for a specific criterion or subcriterion. However, if the variation is greater than 30%, then a discussion, agreement, and rescoring have to be undertaken. If this is the case, then the same criterion applies after the rescoring (e.g., in less than 30% variation the scores are averaged). If after the rescoring a less than 30% variation is not achieved, then the team leader must take the best view and complete the consensus scorebook. We suggest adopting and following this simple set of consensus criteria established by the EQA assessment in order to ensure a fast and efficient, but still objective, consensus process.

In some instances, further clarification may be needed before undertaking the scoring or rescoring; if this is the case, then one or more site visits may need to be arranged. Site visits are a normal part of the self-assessment process when an organization is applying for an award. This is because there is normally a significant time lapse between the preparation of the submission and its subsequent assessment. However, if the submission is being done for self-assessment purposes only, site visits are only required if further clarification is needed to support either the scoring or rescoring process.

4.3.1.7 Stage 7: Producing the Feedback Report

Once a consensus has been reached, the following stage consists of the assessment team's leader writing a first draft feedback report, which must later be circulated to the other members of the assessment team. In this case, the assessment team members have to review the report and include any observations or comments or make any amendments they believe should be incorporated into the report. The feedback report will be the major outcome of the self-assessment process. In particular, Porter and Tanner (1998) suggest that a well-written and structured feedback report provides the following information:

- *An overview of the assessment process.* This might include how it was conducted, who participated in the assessment, the criteria and subcriteria considered and evaluated, how the data were collected, etc.
- *An executive summary.* This should provide a concise description and impression of the assessment and submission.
- A list of strengths and opportunities for improvement for each criterion and subcriterion.
- The overall and individual score for each criterion and subcriterion.

Finally, the self-assessment report should be passed on to the review committee for review and analysis. The review committee will then discuss and coordinate improvement plans and actions, and their prioritization, with top management. It is typically at this stage that the assessment team concludes the self-assessment exercise, although the review committee may still require further clarification from either the team leader or the whole assessment team. We suggest that top management and the review committee include the assessment team in the following stage of the QMS diagnostic, in this case, the quality auditing process. The inclusion of the assessment team in the proposal and implementation of the appropriate measures undertaken to bridge the gaps between the BEM criteria and the organization's current performance is also recommended. The self-assessment team would be comprised of employees who are "experts" and have an in-depth knowledge of the organization's functioning and processes. For this reason, their participation can prove invaluable to the successful completion of the post-self-assessment stages.

4.4 Quality Management Audits

For some organizations, quality audits are a mandatory activity that needs to be performed in order to comply with requirements from their customers, suppliers, partners, collaborators, or the industry sector, and even to fulfill government regulations. Quality audits help organizations, and those that request them, monitor and assure that a QMS is in place and working effectively. In turn, products or services that comply or exceed quality standards would be expected. Professor Oakland (1989) comments, "A good quality system will not function without adequate audits and reviews." It is for these reasons that we suggest, as part of the QMS and business processes diagnostic methodology proposed in this chapter, the institution of quality audits. In this way, quality audits will provide further information about the QMS and organization's business processes, particularly whether they comply with the required standards. We have to clarify that it is not within the scope of this section to provide a detailed review of the quality auditing process. This is an extensive topic within the QM area that has been clearly and extensively covered in, for example, specialized books by Mills (1993) and Arter (2003). Rather, the main objective of this section is to explain how quality audits can be integrated and contribute to the diagnostic of the status of a QMS and business processes. Figure 4.3 illustrates this.

In general terms, quality audits fall under three main categories: first-party audits, second-party audits, and third-party audits. In a first-party audit, the assessment of the quality system against a particular standard is carried out internally within the organization, while in a second-party audit, it is done by a customer or supplier. In a third-party audit, an independent organization not involved in any contract with the customer and supplier, but acceptable to both of them, carries out the audit. We consider that a first-party audit is the easiest and most efficient type of audit to perform when this activity is integrated into the QMS and business processes diagnostic. This is because the same team involved in the maturity assessment and self-assessment process can conduct the quality audit. As this team may have been involved from the initial stage of defining the maturity of the QMS and through the self-assessment process, it would already have an in-depth knowledge of the QMS and core business processes of the organization. In addition, by the end of the quality auditing process, the assessment team members would have acquired an overall picture of the status of the organization's QMS and business processes. This will also facilitate the reporting and debriefing of such status to top management.

Figure 4.4 presents a general illustration of the stages of a quality audit process. In the initial planning stage, different aspects that include the audit's purpose, timelines, scope, resources needed, etc., are identified and defined. Once the audit plan is complete, its implementation can begin. The implementation stage consists of several activities that include the collection of information, its comparison against the standard or criteria, and the initial

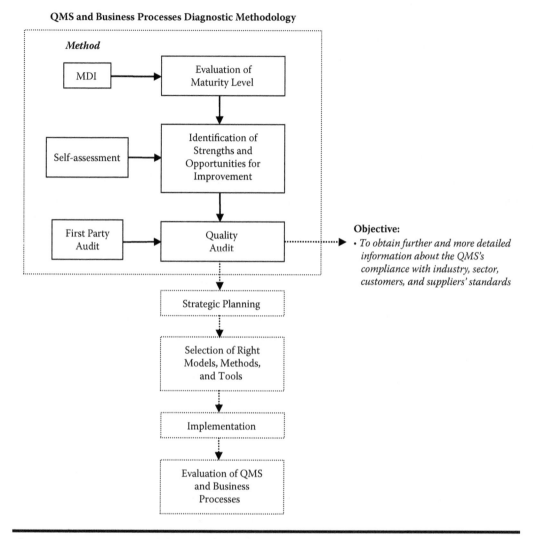

Figure 4.3 Overview of the QMS diagnostic methodology and role of quality audits.

review of this comparison. In terms of the collection of data, quantifiable evidence is more reliable than subjective evidence, so auditors must aim at collecting this type of information whenever possible. The selection of the most appropriate method for collecting data should be based on an evaluation of cost, time, the risk of obtaining a bad judgment, and the resources available to perform the audit.

The initial review stage follows the data collection activity. As part of this activity, the auditors review and analyze the data obtained after their comparison against the standard. This will lead to the allocation of nonconformities. Finally, the auditors will prepare a report and debrief the organization on the differences found between the evidence collected and the standard.

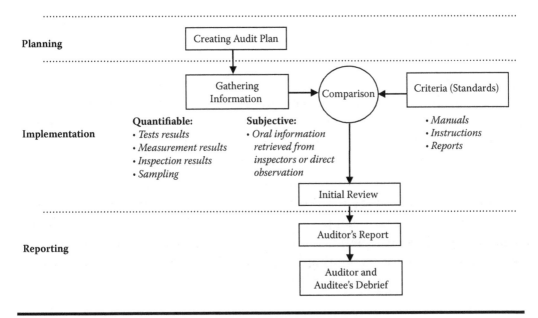

Figure 4.4   Quality auditing process.

As the quality auditing process will provide an in-depth review and evaluation of an organization's QMS against a specific standard, the information obtained from it will enrich the overall diagnostic of the QMS and its business processes.

4.5 Role and Importance of the QMS and Business Processes Diagnostic on Operational Improvement and Business Strategy

A vital and initial step that will enable an organization to select, design, implement, or improve a QMS is to diagnose and understand the maturity of its QMS and the strengths and weaknesses of its core business processes. Evaluating whether the QMS complies with the standards set by the organization's customers, suppliers, partners, etc., is also part of this initial step. Once achieved, the organization can then propose and deploy an action plan to address the areas for improvement highlighted in the overall diagnostic of its QMS and business improvement activities. In the particular case of self-assessment processes, empirical evidence and our experience reveal that the decisions and improvement agenda created based on such processes are rarely documented. As a result, it is difficult to estimate the degree to which self-assessment influences improvement actions or whether or not its results are simply kept in the desk drawer of the organization's CEO or directors, with no improvement actions being drawn and implemented. It is therefore of major importance that the organization integrate the diagnostic of its QMS and business processes into its business plan and strategy. This would provide the organization with an effective mechanism by which to (1) define adequate improvement actions; (2) transform these improvement actions into an improvement agenda; (3) implement, review, and sustain the improvements; and (4) document the results obtained. Recent research by Rocha-Lona et al. (2010) suggests that BEMs are suitable frameworks for supporting strategic planning and business improvements. Similarly, the diagnostic of a QMS can also support improvement actions if integrated into the organization's business plan and strategy. In subsequent chapters, we propose a framework for integrating the results of the QMS diagnostic into an organization's business plan and strategy.

4.6 Summary

In this chapter we have highlighted the importance of diagnosing the current status of the QMS implemented in the organization, and its business processes, and have provided a methodology for performing such a diagnostic. In particular, the initial step of the diagnostic consists of defining the maturity level of the organization's QMS. To do this, we have proposed an MDI developed and adapted from the six-level categorization model of Dale and Lascelles (1997). Defining an organization's QMS maturity will provide not only a better understanding of its quality capabilities, structure, procedures, and processes, but also a comparative platform from which to later assess any improvements achieved.

As a second step, the QMS and business processes diagnostic methodology we propose suggests that an identification of the strengths and opportunities for improvement in the organization's business processes has to be carried out. To do this, a self-assessment exercise using a main BEM, or alternatively a tailored model that draws different criteria from different BEMs, is recommended. For this reason, we have taken a detailed look at the key steps in the self-assessment exercise and proposed a series of stages based on the best practices of experts in the area, the literature, and our own experience. These include setting the organizational environment for the self-assessment process, selecting a BEM, forming and training the assessment team, collecting the data and information needed for the self-assessment process, achieving consensus, assessing and scoring, and producing a feedback report. Understanding and practicing these steps are vital to performing a self-assessment exercise and developing the organization's capability to carry out such processes. Finally, the methodology also integrates quality audits as a means to providing information about the compliance of the QMS in relation to customers, suppliers, industry, or government standards.

In this chapter we have also briefly discussed the importance of integrating the QMS diagnostic into the organization's plan and strategy as an approach to more effectively drive improvement actions and their implementation. The following chapters cover this issue in more detail.

4.6.1 Key Points to Remember

- It is essential for organizations to understand the current status of their quality structure, procedures, processes, and resources to enable an effective selection, design, implementation, or improvement of a QMS.
- This understanding can be obtained by diagnosing the maturity of their QMS, the strengths and weaknesses of their core business processes, and whether their quality procedures comply with the required standards.
- The organization's QMS maturity can be diagnosed using the MDI we propose in this chapter.
- The strengths and weaknesses of the organization's core business processes can be diagnosed by performing a self-assessment process using a BEM and following the steps also presented in this chapter.
- Compliance with standards is recommended to be diagnosed through a first-party audit and following the steps we have presented in this chapter.
- It is of major importance for an organization to integrate the diagnostic of its QMS and business processes into its business plan and strategy.

References

Antony, J., and Preece, D. (2002). *Understanding, managing and implementing quality: Frameworks, techniques and cases*. Routledge, London.

Arter, D. R. (2003). *Quality audits for improved performance*, 3rd ed. ASQ Quality Press, Milwaukee.

Dale, B. G., and Lascelles, D. M. (1997). Total quality management adoption: Revisiting the levels. *TQM Magazine*, Vol. 9, No. 6, pp. 418–428.

Gadd, K. W. (1995). Business self-assessment: A strategic tool for building process robustness and achieving integrated management. *Business Process Re-engineering and Management Journal*, Vol. 1, No. 3, pp. 66–85.

Hillman, G. P. (1994). Making self-assessment successful. *TQM Magazine*, Vol. 6, No. 3, pp. 29–31.

Mills, D. (1993). *Quality auditing*. Chapman & Hall, Cornwall, UK.

Oakland, J. S. (1989). *Total quality management*. Butterworth-Heinemann, Oxford.

Porter, L. J., and Tanner, S. J. (1998). *Assessing business excellence*. Butterworth-Heinemann, Woburn, MA.

Rocha-Lona, L., Torres, A. D., Garza-Reyes, J. A., Soriano-Meier, H., and Salinas-Navarro, D. (2010). The use and impact of business excellence models on organisations: The case of the Mexican quality award. At Proceedings of the 20th International Conference on Flexible Automation and Intelligent Manufacturing (FAIM), San Francisco, July 12–14.

Further Suggested Reading

Dale, B. G., and Lightburn, K. L. (1992). Continuous quality improvement: Why some organisations lack commitment. *International Journal of Production Economics*, Vol. 27, No. 1, pp. 57–67.

Lascelles, D. M., and Dale, B. G. (1991). Levelling out the future. *TQM Magazine*, Vol. 3, No. 2, pp. 125–128.

Chapter 5

Strategic Quality Planning

5.1 Introduction

Strategy is a term that all business people believe they know and understand (O'Regan and Ghobadian, 2002). When speaking about management concepts such as Total Quality Management or business strategy, strategy has different connotations, and there is little agreement in terms of defining what it is, despite decades of development. It is therefore very important to provide a definition that suits the purpose of this section of the book, which is related to strategic quality planning (SQP). The concept of a strategy has its origins in the military, perhaps with Sun Tzu's introduction to the *The Art of the War*, written in 551 B.C. (Tzu, 2001). In this context, strategy refers to an army's ability to use its available resources to defeat enemies. This same idea can be extrapolated to business activities; however, its use in this context refers to the ability of business organizations to outperform their rivals in a competitive market.

A number of definitions of the concept of strategy, within a business context, have emerged in response to the evolution of business activity. Mintzberg (1994) defines strategy as a plan or a guide for action in the future. For Porter (1996), strategy is a set of activities that create a valuable position, which differentiates an organization from its rivals. In terms of industry analysis, Oliver (1996) conceives strategy as having an understanding of a particular industry, and determining the organization's position in it. All these definitions are correct, and can be used in several ways, depending on the business context. This book considers strategy in the sense of a plan; in particular, it considers strategy as a structured process in which organizations can define their course of action for the medium and long term. This set of plans, along with the process to produce the plans, forms what is actually the strategic planning process. Therefore, the strategic quality plan is developed under this framework, with the aim to provide a robust quality management system that raises the organization's level of quality to a high degree.

5.2 Strategic Decision Making—Why Does It Matter?

When developing a strategic quality plan, it is useful to understand and select the right quality models, methods, and tools, since they are decisions that have to be cost-effective, and ultimately support the business performance. In this way, *strategic decisions* are concerned with an organization's long-term direction, and are value oriented, requiring top management's involvement. These decisions are usually made by the board of directors, and involve issues such as opening new facilities and investing in resources such as a quality management system (QMS). *Tactical decisions* deal with the implementation of strategies and plans for particular functions or business areas. They are usually made for the medium term to support the overall strategy on specific issues, and are made by the head of the business unit. Finally, *operational decisions* are concerned with resources, processes, people, and their skills, on a day-to-day basis, and are employed to reach short-term targets (Figure 5.1).

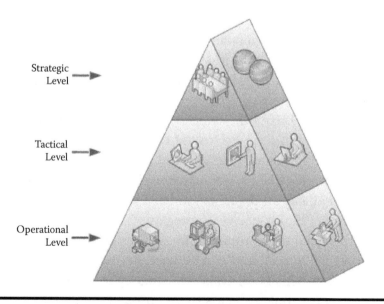

Strategic Level →

Tactical Level →

Operational Level →

Figure 5.1 Levels of decision making.

It is widely recognized that strategic decisions are usually made at the corporate and business levels in organizations. These decisions typically have a strong influence on operations, and affect all of the organization's activities. An important characteristic of strategic decisions is that they are usually supported by top management at the highest levels (Harrison and Pelletier, 2001). Identifying and defining the strategic divide between strategic, tactical, and operational decisions is important for prioritizing plans and actions (Table 5.1).

Table 5.1   Strategic, Tactical, and Operational Decisions

	Strategic	*Tactical*	*Operational*
Focus of decision	Setting objectives and vision for the QMS Achieving sustainable competitive advantage	Designing the QMS by selecting quality models, method, and tools Key implementations of quality management strategies	Implementing and monitoring the quality methods and tools Day-to-day operations Process documentation
Level of decision making	Senior management, board of directors	Heads of business units Director of operational excellence	Supervisory
Scope	Whole organization	Business area or functional area (e.g., production)	Department
Time horizon	Long term (years)	Medium term (months to years)	Short term (days, weeks, months)
Certainty/ uncertainty	High uncertainty	Some uncertainty	High certainty
Complexity	Highly complex	Moderately complex	Comparatively simple
Examples	Decision to implement the EFQM business excellence model in the entire organization	Decision to apply Six Sigma and Lean in core processes along with some certifications in key areas/products based on ISO standards	Using process mapping to capture the voice of customers Applying SPC in the customer services department to monitor complaints

The decision to set and deploy a QMS is strategic, and requires a considerable investment in capital, time, and human resources that will affect the organization in the medium and long term. Senior management and directors will have to deal with the decisions to set the QMS. Later on, they will also have to choose the right quality methods and tools to translate quality strategies at the tactical and operational levels, and ultimately to deliver benefits. Failure to make a good decision in these issues can result in huge pitfalls that can severely affect business performance and finances.

5.3 Strategic Quality Planning Model for the QMS

SQP is the set of decisions and actions that result in the formulation and implementation of quality programs to achieve improvement objectives. SQP has become essential to provide direction to quality management efforts and continuous business improvement. It is composed of four stages: business analysis, strategy formulation, strategy implementation, and evaluation (Figure 5.2). This process is the result of the evolution from basic financial planning to a formalized process that involves the internal and external analysis of the organization, the generation of plans and objectives, the implementation of those plans, and the constant monitoring of outcomes. Strategic planning focuses on the direction of the organization in the long term, and considers the necessary actions to improve its performance. This approach, usually called prescriptive or deliberate, seeks to match the organizational

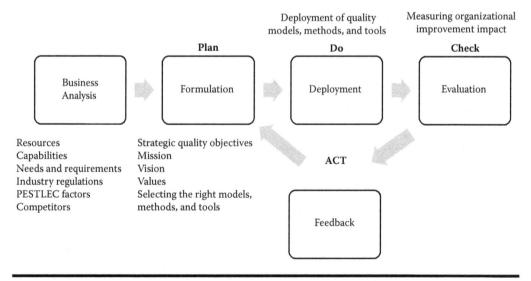

Figure 5.2 Strategic quality planning model for the QMS.

strategy with the environment in which it operates (Campbell et al., 2002). Although there are some criticisms of this approach, such as its lack of flexibility to cope with unexpected changes, its effectiveness has been tested for setting long-term objectives and formulating policies and plans. In fact, SQP has the potential to positively affect organizational performance and, furthermore, to be valuable in unstable environments.

Like any other approach, strategic planning has its drawbacks, such as its lack of flexibility to cope with unexpected changes (Mankins, 2004). Mintzberg (1994) suggests that the most successful strategies are visions, not plans. However, he also recognizes that organizations must plan to coordinate their activities, prepare for the future, and control their operations. Ultimately, visions mean little if they are not methodically put into practice, in a formalized or informal plan. Additionally, it is not enough to identify industry trends, or in which business an organization should be (Oliver, 1996), because "words" and "visions" need to be translated into actions, and actions must be supported by a set of coherent decision-making and planning activities. Thus, SQP, whatever its conception and source, enables the achievement of targets in the medium and long term, which in turn facilitates and supports the leaders' vision for the organization. In the following section, we describe the SQP model for the QMS.

5.3.1 First Stage: Business Analysis

Business analysis is concerned with monitoring, evaluating, and disseminating information drawn from the environment in which the organization functions. This process comprises internal and external analysis, and helps to identify the factors that support the SQP process. We suggest that the internal analysis comprises resource analysis (i.e., human and capital), and analysis of the organization's needs and capabilities. A SWOT analysis can be very helpful for this purpose. A further element that must be determined is the quality maturity level of the organization, which we covered in Chapter 4. This also should be complemented with the self-assessment approach, also covered in Chapter 4, in order to identify strengths and opportunities based on the BEM criteria. This is the core internal analysis, which, along with the setting of the organization's needs and requirements, will shape strategic objectives.

In addition, internal intelligence has to complement the previous analyses. Internal intelligence refers to the activity of structuring information to build the strategic quality plan and business strategy. This analysis can cover areas such as business models, systems thinking, quality models, knowledge management, IT, resource and development, and patents, among many others. This will simply depend on the nature of the business. This analysis will help to identify the best practices worldwide, by the best companies and research centers. There are some sources of information, such as those presented in Table 5.2, which can help to support this stage.

On the other hand, external analysis comprises the analysis of political, economic, sociodemographic, technological, legal, environmental, and cultural (PESTLEC) factors. This analysis should provide information related to competitors (i.e., benchmarking), financial analysis, country intelligence, industry tendencies, etc. To start this task, it is first necessary to define

Table 5.2   Databases and Models Containing Information Valuable to Business Management

Database	Service	Coverage
ABI Inform	Covers more than 1,600 leading business and management publications.	Worldwide
Business Source Premier (EBSCO)	Includes full text for more than 1,125 business publications. It provides expanded indexing and abstracts for some businesses.	Worldwide
Expanded Academic ASAP (info tract)	Covers a range of scholarly journals for a wide range of academic disciplines, including business and management.	Worldwide
Sage Management and Organization Studies	Covers publications in the areas of business and management, including organization studies, human relations, marketing, etc.	Worldwide
Emerald	Includes over 2,000 full-text business and management journals.	Worldwide
SWOT Analysis	Tool that helps to identify strengths, weaknesses, opportunities, and threats.	Generic
JICA model	Diagnostic tool to identify areas of improvements designed for small and medium-sized enterprises (SMEs).	Generic
Self-assessments using a BEM such as the EFQM model or Baldrige model	Powerful tool based on self-assessments to identify areas of improvements (when used for such purpose). See Chapter 2, Section 2.3, and Chapter 4, Section 4.3.	Generic

what sort of information the organization needs, and to determine areas that require this information (i.e., business divisions, marketing department, financial department, quality department, R&D department). Second, it is necessary to allocate resources to get the information, including the IT resources to process and structure the information. Finally, given these constraints, organizations should consider finding external consultancy, although it is recommended that they keep overall control of the business intelligence process. Once these points are clear, a business intelligence framework can be tailored to cover specific needs. Table 5.3 provides a sample of databases that provide business intelligence of several industries, products, and markets. It is very useful if organizations can have some of these specific

Table 5.3   Databases Covering PESTLEC Factors

Database	Service and Industries	Coverage
Amadeus	Amadeus is a comprehensive database containing financial information on approximately 9 million public and private companies in 38 European countries.	Europe
Orbis	Orbis is a global database that has financial information on over 35 million companies.	Worldwide
Osiris	Osiris is a comprehensive database of listed companies, banks, and insurance companies.	Worldwide
Compustat	Compustat is a North American database that allows financial analysis of major U.S. and Canadian companies. Other resources also provide information from all non-North American companies (Compustat global).	Worldwide
Global Insights	Global Insights provides country intelligence in 200 countries and more than 170 industry analyses.	Worldwide
Global Market Information Database (Euromonitor)	The Global Market Information Database provides current and forecasted economic indicators (including GDP, banking, government expenditure, consumer expenditure/prices, disposable income) for over 200 countries.	Worldwide
OECD Economic Outlook	Provides a comprehensive statistic data of the of the 30 OECD economies.	Worldwide

Table 5.3 *(Continued)*   Databases Covering PESTLEC Factors

Database	Service and Industries	Coverage
Economist Intelligence Unit	The Economist Intelligence Unit provides information and forecasts on more than 200 countries and 8 key industries.	Worldwide
Factiva	Factiva provides business news and information along with international stock exchange indices from quoted companies.	Worldwide
Mintel	Mintel database provides UK consumer market research and related trade. The reports cover standard markets, essentials, food and drink, leisure, pursuits, catering, travel, consumer retail markets, financial markets, and products.	UK
Keynote	The Keynote database covers over 250 titles in approximately 25 consumer, business, and industry sectors for the UK.	UK
Global Best Practice	The Global Best Practice is knowledge resource for best practices, benchmarking, business risks, and controls.	Worldwide
BPIR	BPIR is a business performance resource that provides a range of information of best practices for improvement and benchmarking purposes.	Worldwide
Excellence One	Excellence One provides a comprehensive database of best practices of quality management, articles, cases studies, and insights from successful organizations.	Europe

services in order to conduct business intelligence that can meet for several purposes. These might include the building of the strategic quality plan and business strategy, among others.

Before formulating strategic quality plans, organizations must carry out an exhaustive assessment of internal and external factors. Furthermore, leaders must recognize what strategic factors are relevant to their cause, so they can adopt an adequate or proper strategic analysis factor framework. The literature offers a wide range of techniques to carry out internal and external analysis, and it is the responsibility of top management and strategists to select the appropriate tools, and prioritize this strategic analysis based on their own requirements.

5.3.2 Second Stage: Strategy Formulation— *Objectives, Mission, Vision, and Values*

Strategy formulation is commonly referred to as long-range planning, and is concerned with developing an organization's mission, vision, objectives, strategies, plans, policies, and values. This stage should involve executives in defining the business the firm is in, its objectives, and the means it will use to accomplish those aims. This process is one of the core areas of the formalized strategic quality planning process and, according to Hewlett (1999), aims to create a company's mission statement, and then to translate this statement into goals with specific time frames and measurable results. The aim of the formulation process for the QMS is concerned with business goals, quality management programs, and most importantly, the selection of the business quality models, methods, and tools. The formulation of strategies, as suggested by Weelen and Hunger (2002), is not one of the main aims in this process. It is frequently assumed that strategies or ways of doing things are part of this process; however, strategic quality planning is different from strategic thinking (Mintzberg, 1994). Strategic formulation is based on an analysis of internal and external factors to establish clear objectives and plans to reach specific business goals. On the other hand, the selection of strategies or the creation of one is not concerned with analysis, but with the synthesis that is achieved through strategic thinking. Thus, for the purpose of this book, the formulation stage matches the *plan* of the PDCA Deming cycle, and is concerned with the setting of the mission, vision, objectives, and plans, as shown in Figure 5.2.

5.3.3 Third Stage: Deployment—*Putting Plans into Action*

Strategy deployment refers to the group of activities, plans, and programs put into practice (Weelen and Hunger, 2002). The strategy deployment process is the critical stage at which the business plans must deliver results. It is frequently argued that the success of the strategic quality planning process relies on how well the plans are translated into actions. This means how well managers execute the plans and translate models, methods, and tools into effective operating terms (Karplan and Norton, 2001). However, even a good plan does not guarantee the desired outcomes; Sterling (2003) comments that "effective deployment of an average plan beats mediocre deployment of a great plan." It is necessary that managers strongly commit to the deployment process, as it is frequently argued that little effort is made to translate plans into actions, after investing lots of capital and human resources in previous stages (Allio, 2005). Thus, effective deployment depends upon a combination of several factors, which include good "planning in deployment," appropriate training at all levels, and a complex framework able to link strategic objectives and plans with operational actions and measures.

5.3.4 Fourth Stage: Evaluation and Control—Setting the Metrics to Measure Performance

This stage refers to monitoring key performance indicators (KPIs) and, in some cases, taking action at the planning and deployment stages to reach business goals. Measuring organizational improvement is not an easy task, and managers should establish performance in terms of return on investments, quality costing, and operational and tacit benefits, such as productivity measures. Chapter 3 has addressed the importance of having a process-centered organization, which in turn should facilitate the establishment of measures in terms of the business value-added processes. Based on our experience, some improvement projects can provide factual results after several months or even years. However, when applying tools such as statistical process control (SPC) charts, single-minute exchange of die (SMED), Pareto analysis, 5S*, and others, the results should be seen in weeks or months at the latest. It is very important to set the right time frames for the improvement projects at the planning stage, to facilitate their evaluation in terms of the expected outcomes. In this way, the feedback regarding organizational improvements should flow between the current performance results and the objectives set at the planning stage, in order to modify improvement strategies or the way of deploying them. Thus, this critical stage relates organizational performance and the QMS aims to determine whether an organization is achieving its strategic objectives or not.

5.4 Using SQP in a Pharmaceutical Company

This section aims to provide an example for setting a strategic quality plan based on SQP concepts, and complemented with Akao's model (currently a hoshin kanri methodology approach) (Akao, 2004). The quality plan was designed for an operation site of one of the top 10 multinational

---

* 5s is the method that uses five Japanese words, each beginning with an "s": *seiri, seiton, seiso, seiketsu*, and *shitsuke*. 5s refers to the way an organization organizes its workspace for efficiency and effectiveness. The method consists of identifying and storing the items used, maintaining in order the area and items, and then sustaining the new order for efficiency and effectiveness.

pharmaceutical companies. The company has been recognized as one of the industry leaders worldwide, and holds several international quality awards. However, despite the high-quality maturity level of this organization, the fact that the pharmaceutical industry is constantly changing, in terms of regulations, requires detailed strategic quality planning that addresses those changing needs and demands from all stakeholders.

The strategic quality plan for this company is divided into four stages: (1) a five-year vision, (2) the one-year-plan, (3) deployment across divisions, and (4) monthly-annual evaluations.

1. *Five-year vision*: King (1989) suggests that the five-year vision includes a draft plan by the president and executive group, which will enable them to develop a revised vision they know will produce desired results and outcomes in the long term. In order to generate a realistic and achievable vision for the pharmaceutical operation site (POS), it was necessary to consider the internal and external factors, primarily the main barriers, resources, and capabilities. In this way, a five-year vision was established with the aim of becoming the most competitive operation site of the corporation, and with the best high-quality standards. The company currently uses the EFQM model as a general umbrella to plan, deploy, and coordinate all improvement programs in the operation's sites worldwide, and the director of excellence and board of directors are in charge of the coordination of these efforts.

2. *One-year plan*: This involves the selection of specific quality methods and tools that were selected mainly on the basis of feasibility and cost–benefit analysis, for achieving the desired results in the short to medium term. This is expected to support the five-year vision of the POS, and to provide specific results in less than one year. At this stage, Akao's model (Akao, 2004) was very useful, since it helped to structure the detailed strategic quality plan, so as to reach key objectives established for the POS (Table 5.4).

3. *Deployment across divisions* (i.e., departments and business value-added processes): This stage focuses on the identification of key implementation items, and a consideration of how they can systematically support the strategic quality plan. Since the POS has several isolated quality improvement initiatives, it was suggested that all of them be integrated and centrally administered, in order to homogenize strategic objectives and reduce operation costs by sharing resources. This means that when someone at the company identifies an area of opportunity or a weak

Table 5.4 Strategic Quality Plan Based on the Hoshin Kanri Approach for POS

Five-Year Vision		Objective: To become the most competitive operation site of the corporation with the best high-quality standards.									
Key Objective	Responsible Area	Goal		Deployment Strategies	Objectives		Improvement Approach				
		Medium Term	Long Term		Medium Term	Long Term	Quality	Cost	Cycle Time	Safety	
Customer satisfaction	Quality assurance + supply chain management	To deliver products to our customers according to schedules and within quality specifications	To retain existing customers and get new customers based on our high-quality products and services	Delivery cycle time improvement; Customer satisfaction measurement; Follow-up of customer complaints	80% on-time deliveries; To build and deploy customer satisfaction surveys with a trimonthly evaluation frequency; Product complaints < 10%; Complaint solutions < 30 days	98% on-time and on-budget deliveries; Customer satisfaction survey applied to 100% of clients; Product complaints < 5%; Complaint solutions < 10 days	To reduce quality analysis time for each batch; Investigation of root causes of each customer complaint	Cost–benefit analysis of customer satisfaction measurement	To establish "right first time" as a key performance indicator	To establish an environmental health and safety department as a focus area to link programs concerned with environmental care to customer satisfaction	

To get leading and competitive personnel		To get stable and efficient processes	
Human resources + all departments		Production + engineering	
development area	To set a talent management and development for all key employees in the organization	To deploy Lean principles and reproduce them in selected business value-added processes	To get new portfolios of products to be manufactured on-site
	To get career planning and development resources of the different parts of the organization	To increase production volume of the current products	To improve overall productivity (OP)
Align the objectives and resources of the workforce during the first month of the current year	Cascade objectives down into the workforce during the first month of month before the end of the year	To reduce manufacturing throughput time (TPT)	To increase capacity utilization (CapUT)
Manage workforce performance annually	Cascade objectives down into the workforce 1 month before the end of year	TPT < 110 days	OP ≥ 63%
Manage workforce performance twice per year		CapUT ≥ 35%	
		TPT < 90 days	OP ≥ 80%
		CapUT ≥ 55%	

To get leading and competitive personnel	To get stable and efficient processes
Follow up strategic plan and objectives / Giving feedback to subordinates	Process validation and critical parameter standardization
To establish a compensation and benefits program	To identify and eliminate waste
To develop a program to attract, select, train, develop, retain, promote, and move employees	To improve manufacturing process time / To get flexible processes
To participate in programs such as "best place to work" and a nondiscrimination policy program	To develop health and safety programs such as those recommended by the Occupational Safety and Health Administration (OSHA)

process, attending to it is prioritized on the basis of the strategic quality plan and the alignment with the company's business strategies. In this way, the director of excellence can schedule the improvement issue and systematically make an assessment, and then select the best methods and tools to make the improvements as soon as possible.

Management at the POS decided to deploy, in key business value-added processes, the Six Sigma improvement method, based on the define, measure, analyze, improve, control (DMAIC) process. The literature and consulting firms offer a wide range of programs to deploy such an improvement method, and there are many key issues that have made this approach very successful. This includes the robust training system, the hard data collected in business processes, and the strong statistical background of its tools, among others. Goldstain (2001) also suggests that strict project reviews and evaluations are among the key tools to succeed in improvement methodologies, such as Six Sigma. He suggests selecting improvement projects based on key processes and considering the following issues:

■ Ensure active participation of senior executives. The involvement of senior executives is essential to achieve the company's vision. Senior managers have to be present at new project launches in order to motivate participants and commit them to the project. They have to effectively communicate the importance of people's involvement, and the key objectives that are expected. These factors are explored in more detail in Chapter 7, where we have defined them as critical success factors (CSFs) for the effective design, implementation, or improvement of QMSs or to successfully carry out improvement initiatives.

■ Make it relevant for managers. Since the human resources for the improvement projects will come from current employees, managers should share human resources, and you need to make sure they understand the common improvement objectives for the whole business.

■ Make it relevant for people. Team members assigned to develop an improvement project should be engaged with and committed to the project. Make sure they understand the win-to-win philosophy, instead of perceiving such projects as extra work or useless activities. We suggest organizations to achieve this by considering participants' profiles, interest, skills, capabilities, certifications, and career plans.

Detailed deployment. Table 5.4 provides the strategic quality plan to become the most competitive operation site of the corporation with the best high-quality standards. It first focuses on key objectives, such as customer satisfaction, stable efficient processes, leading and competitive personnel, and getting the lowest cost per unit. Then, the quality plan provides the areas of responsibility, along with their specific goals and the improvement strategies to be deployed. Finally, it provides the medium- and long-term objectives, with specific metrics to evaluate these based on quality, cost, cycle times, and safety, which are of high priority for the POS. It is believed that improvement of these issues will directly impact on financial performance and industry regulations.

Then, by way of example, Table 5.5 provides the action plan for one of the key objectives: getting stable and efficient processes. This involves the improvement quality method used to achieve it, and the targets or metrics used to determine when the objective has been met. Here, the company uses the Six Sigma method and sets the target to get an overall productivity of ≥63%. The deployment also covered the following: (1) the identification of customer needs, (2) the development of the action plan for the project portfolio management, (3) the selection of personnel with competitive skills and an interest in developing the improvement project, and finally, (4) the training (certification) for those who were selected in order to provide them with the knowledge and skills to develop the project. This last point is also related to the goal of "to get leading and competitive personnel."

4. *Monthly-annual evaluations*: This is related to the analysis of things that helped or hindered progress, and the activities that will benefit from any lessons learned. In this stage, a balanced scorecard approach was suggested to develop KPIs, which would be measured and evaluated monthly. Finally, an annual diagnosis and meeting were also suggested, with all staff involved, and with the objective of reviewing progress, and setting corrective actions and business improvement strategies for current and future projects.

5.4.1 Cost–Benefit and Non-Cost–Benefit Analysis

Table 5.6 shows the cost–benefit analysis for implementing the strategic quality plan. In order to estimate the costs, quotes were obtained from suppliers available in the country where the POS is located. In other cases, quotes came from similar companies in different countries offering the services. The financial benefit was calculated according to savings generated by previous improvement initiatives or similar actions reported by the

Table 5.5   Action Plan for Pharmaceutical Operation Site

Hoshin objective title:	To become the most competitive operation site of the corporation with the best high-quality standards	Management:	Production + engineering
Department:	Human resources	Approved date:	February 2012
Review team:	All managers	Next review:	July 2012
Current external status:		The pharmaceutical industry is an important and profitable business around the world. Companies immersed in this industry must develop manufacturing strategies that give them competitive advantages. In this context, they should have the ability to manufacture high-quality products at lower costs than competitors, using some of the available business process improvement methodologies.	
Key Objective		*Strategy*	*Objective*
To achieve stable and efficient processes		To reduce manufacturing throughput time	Throughput < 110 days
Goals			
Medium term	To increase the production volume of the current products	To improve overall productivity	Overall productivity ≥ 63%
Long term	To get new portfolios of products and have them manufactured on the site	To increase capacity utilization	Capacity utilization ≥ 35%

POS in the last year. Based on this analysis, and considering the net present value of profits, it was possible to conclude that the implementation of the strategic quality plan is acceptable, since it provides a cost–benefit ratio of 2.7. The nonfinancial benefits are primarily related to the understanding of the root causes of the main quality problems, the identification of customers' needs and requirements, the development of employees, the benefits of

Table 5.6 Cost–Benefit Analysis of the Strategic Quality Plan

Key Objective	Improvement Approach (Activities)	Cost (USD)	Benefit (USD)
Customer satisfaction	To reduce quality analysis time for each batch		
	80 hours for 30 persons[a]	$46,000.00	$270,153.85
	Investigation of root cause of each customer complaint		
	Specialist of complaint system	$23,076.92	
	Complaint staff (phone number + Internet + assistant)	$10,830.00	
	Customer satisfaction measurement		
	Generation and application of the survey to all customers[b]	$8,306.77	
	Results analysis and conclusion report (marketing administration)	$27,692.31	
To get stable and efficient processes	Process validation and critical parameter standardization		
	Validation area (1 manager + 5 engineers)	$110,769.23	
	To identify and eliminate wastes in key processes	c	$276,923.08
	To identify and eliminate wastes	c	$1,523,076.92
	To develop health and safety programs		
	Safety, health, and environmental area (1 manager + 3 specialists)	$87,692.31	
	Occupational safety and health administration program		
	Hazard disposal programs	c	$15,384.62

Table 5.6 (Continued) Cost–Benefit Analysis of the Strategic Quality Plan

Key Objective	Improvement Approach (Activities)	Cost (USD)	Benefit (USD)
To get leading and competitive personnel	Objective plan assessment and feedback to subordinates		
	Software for employee assessment[d]	$53,846.15	
	Training in effective communication skills for managers[e]	$76,923.08	
	To establish a compensation and benefit program		
	Benefit and compensation software[f]	$23,076.92	
	Talent management		
	Talent management software	$27,083.33	
Lowest cost per unit	To develop at least one local supplier for each imported raw material		
	Supplier development plan + auditing program to them[g]	$153,846.15	
	Equipment maintenance program		
	Maintenance area (1 supervisor + 15 mechanics)	$124,615.38	
	Total	**$773,758.56**	**$2,085,538.46**
	C/B = 2,085,538.46/773,758.56		**= 2.7**

Source: With information of the POS.

[a] Six Sigma training + benefits of implemented projects during last year.
[b] Platinum program of survey company.
[c] Master Black Belt coach + man-hour cost of 30 trainees.
[d] Performance appraisal software.
[e] Consultants services.
[f] HR management software.
[g] Purchasing and lead auditor man-hour costs based on last year's supplier development costs.

avoiding serious accidents, and health benefits. Therefore, it can be concluded that deployment of the strategic quality plan can provide benefits in financial performance, as well as nonfinancial issues related to quality issues and personal development and satisfaction.

5.5 Summary

This chapter deals with SQP and its application to support the QMS. It provides the elements to set a strategic quality plan for medium- and long-term time frames, emphasizing the use of the best decision-making practices. It then reviews some concepts of long-range planning, which properly deployed can help directors and their organizations to achieve their business objectives. The chapter provides a practical guide to produce and implement a strategic quality plan, considering internal/external analysis, strategy formulation, deployment, and evaluation. It then also provides an example of a pharmaceutical company for which a strategic quality plan was developed, based on the hoshin kanri methodology. It is expected that professionals are in a position to generate this type of plan, and can adjust it in terms of an organization's particular requirements, capabilities, and resources.

5.5.1 Key Points to Remember

- Make sure your organization understands strategic quality planning and its main stages.
- Make sure that all staff know and understand the key objectives, values, and mission of the organization. Share your vision with the others.
- Assess your capabilities and resources when setting the strategic quality plan.
- Provide your organization with key business intelligence and infrastructure that help to identify industry regulations and all PESTLEC factors.
- Involve senior managers and directors in improvement efforts.
- Deploy your strategic quality plan with discipline.

References

Akao, Y. (2004). *Hoshin kanri: Policy deployment for successful TQM.* Taylor and Francis Group, CRC Press, Boca Raton, FL.

Allio, M. (2005). A short practical guide to implementing strategy. *Journal of Business Strategy*, Vol. 26, No. 4, pp. 12–21.

Campbell, D., Stonehouse, G., and Houston, B. (2002). *Business strategy.* Butterworth-Heinemann, Oxford.

Goldstain, M. (2001). Six Sigma program success factors. *Six Sigma Forum Magazine*, Vol. 1, No. 1, pp. 36–45.

Harrison, E. F., and Pelletier, M. A. (2001). Revisiting strategy decision success. *Management Decision*, Vol. 39, No. 3, pp. 169–180.

Hewlett, C. A. (1999). Strategic planning for real estate companies. *Journal of Property Management*, Vol. 64, No. 1, pp. 26–29.

Kaplan, R. S., and Norton, D. P. (2001). Building a strategy-focus organization. *Ivey Business Journal*, May/June, pp. 12–19.

King, R. E. (1989). Hoshin planning, the foundation of total quality management. *American Society for Quality Congress*, Vol. 43, No. 0, pp. 476–480.

Mankins, M. C. (2004). Making strategy development matter. *Harvard Management Update*, May, pp. 3–5.

Mintzberg, H. (1994). The fall and rise of strategic planning. *Harvard Business Review*, January–February.

Oliver, R. W. (1996). What is strategy, anyway? *Harvard Business Review*, November–December, pp. 7–10.

O'Regan, N., and Ghobadian, A. (2002). Formal strategic planning. *Business Process Management Journal*, Vol. 8, No. 5, pp. 416–429.

Porter, M. E. (1996). What is strategy? *Harvard Business Review*, November–December, pp. 61–78.

Sterling, J. (2003). Translating strategy into effective implementation: Dispelling the myths and highlighting what works. *Strategy and Leadership*, Vol. 31, No. 3, pp. 27–34.

Tzu, S. (2001). *El arte de la guerra.* Ediciones Coyoacán, Mexico City.

Weelen, T. L., and Hunger, J. D. (2002). *Strategic management and business policy.* Prentice Hall, Upper Saddle River, NJ.

Further Suggested Reading

Have, S. T., Have, W. T., Stevens, F., Elst, M. V. D., and Pol-Coyne, F. (2003). *Key management models.* FT Prentice Hall, Glasgow, UK.

Kaplan, S., and Lamotte, G. (2001). The balance scorecards and the quality programs. *Harvard Business Publishing Newsletters*, March.

Chapter 6

Building the QMS and Business Improvement Plan by Selecting the Right Models, Methods, and Tools

6.1 Introduction

In Chapter 4 we presented a diagnostic methodology that can help an organization perform a thorough evaluation of its quality management system (QMS) and business processes to highlight weak areas that need to be considered for improvement. In Chapter 5 we then provided a method for aiding an organization to align the formulated improvement plans with its strategy and planning. In this chapter we continue with the next step, which consists of selecting the right models, methods, and tools to be adopted to enhance the organization's QMS and execute its improvement plan. The chapter starts by classifying the most popular and widely used business and quality improvement approaches. It then continues with a discussion of the criteria we propose for assessing the suitability of the selected models, methods, and tools to overcome a particular weakness. Finally, a series of steps that organizations can follow in selecting the most appropriate models, methods, and tools is presented.

6.2 Business and Quality Improvement Models, Methods, and Tools: A Classification

During the design, implementation, or improvement of an organization's QMS and business processes, the selection of the right improvement approaches is essential to successfully carrying out the initiative. The last decades have witnessed the development of a large number of philosophies, models, methods, and tools that have been proposed to help organizations in their quest for competitiveness. In particular, business and quality improvement models, methods, and tools play different roles within an organization's QMS and its processes. These roles include

- Providing a philosophy and an approach for business improvement
- Providing a reference for the measure of organizational performance
- Organizing and summarizing the presentation and communication of data
- Providing a structured method for collecting data
- Providing a systematic approach to uncovering root causes and solving problems
- Monitoring and maintaining control
- Prioritizing, implementing, and sustaining improvement initiatives
- Planning
- Investigating and identifying the relationships of process variables

Based on experience, knowledge, and the literature, we provide in Table 6.1 a nonexhaustive summary as well as a structured categorization and alignment of some of the most popular and well-known business and quality improvement models, methods, and tools used by organizations. In this text we refer to models as those nonprescriptive standards that show organizations the criteria or characteristics of business excellence or those required in satisfying their customers' expectations. Examples of models include any business excellence model (BEM) such as European Foundation for Quality Management (EFQM), Malcolm Baldrige, Deming, etc., or any quality management standard such as ISO, British Standards, QS-9000, etc.

On the other hand, we consider methods as those approaches that provide organizations with a philosophy and a "receipt" for improving different aspects of their business operations or products. This category includes main approaches such as Lean manufacturing, Six Sigma, and Total Quality Management (TQM), among others, which explicitly indicate how organizations can improve different aspects of their businesses. To assist in making the selection of the most appropriate models, methods, and tools easier for organizations, we have subdivided our classification of methods into tier 1 and tier 2 methods. Tier 1 methods represent main methods such as Lean manufacturing, Six Sigma, TQM, etc., while tier 2 methods are the pillars that support the main methods by making the achievement of their objectives and implementation possible. For instance, just-in-time (JIT), define, measure, analyze, improve, control (DMAIC), and quality costing are considered tier 2 methods since they complement Lean manufacturing, Six Sigma, and TQM, respectively.

Finally, we have classified as tools those enablers and techniques that support the implementation and operationalization of tier 1 and tier 2 methods. Based on our classification, an example of a tool is *one-piece flow*. In this case, one-piece flow is used as a technique for reducing inventory within the JIT method, which in turn helps reduce waste as part of the Lean manufacturing approach. Similarly, the *Pareto chart* is a tool traditionally used in the define phase of the DMAIC method, which in turn is considered part of Six Sigma.

Table 6.1 offers an organized view of how different models, methods, and tools for business and quality improvement fit and interact with each other. However, arriving at a general consensus with other authors and practitioners on an "all-agreed classification" can prove to be almost impossible. This is because large variations can be found in the perceptions of different authors and the literature. For instance, Professors Barrie Dale and John Oakland present TQM, in their books *Managing Quality* and *TQM*, as the ultimate umbrella from which all quality aspects and initiatives of an organization are initiated. However, Professor Nigel Slack et al. (2006) refer to TQM as an integral part of Lean manufacturing, while Dennis Beecroft (2004) considers it as a stage in the evolution of quality. Similarly, Thomas Pyzdek (2003) presents statistical process control (SPC) as an integral part of Six Sigma and DMAIC, whereas Professor Douglas C. Montgomery (2009) has traditionally presented SPC as a stand-alone quality control and improvement method. Additionally, some tools can be used as part of the implementation and operationalization of different methods; an example of this is the concept of *mistake proofing*. Hagemeyer et al. (2006) consider mistake proofing as having originated from the control phase of DMAIC and Six Sigma. However, mistake proofing is widely referred to in the operations management literature as an essential part of Lean manufacturing, where it is also known as poka-yoke. It is for these reasons that our classification and categorization of operations and quality improvement models, methods, and tools presented in Table 6.1 must be interpreted as a general guide only.

Table 6.1 Most Popular and Well-Known Business and Quality Improvement Models

Models	Tier 1 Methods	Tier 2 Methods	Tools	Objective/Application
BEMs: EFQM, Malcolm Baldrige, Deming, Singapore Quality Award (SQA), Canadian Framework for Business Excellence (CFBE), Australian Business Excellence Framework (ABEF), business performance improvement resource (BPIR), Mexican quality model for competitiveness (MQMC), Shingo, etc. QM Standards: ISO, BS, QS-9000, AS-9100, TS-29001, OHSAS 18001, etc.	TQM Six Sigma	SPC	Control charts (e.g., X bar, R, exponentially weighted moving average—EWNA, p, np), histograms, Pareto charts, defects check sheets, scatter diagrams, inspection sampling, etc.	Process performance monitoring and evaluation
		Quality function deployment (QFD)	The house of quality, questionnaires, surveys, etc.	Interface between customer needs and product features
		Quality costing	Activity-based costing, PAF (prevention, appraisal, failure), quality cost curves, etc.	Effective decision making, setting priorities, and planning
		Benchmarking	Internal benchmarking, competitive benchmarking, functional benchmarking, generic benchmarking, etc.	Comparing results, processes, and procedures against the best of the best
		DMAIC, plan, do, check, act (PDCA), Design for Six Sigma (DFSS)	Checklists, control plans, check sheets, bar charts, tally charts, histograms, graphs, affinity diagrams, systematic diagrams, SIPOC (supplier, inputs, process, outputs, customers) diagrams, brainstorming, flowcharts, cause-and-effect diagrams, scatter diagrams, regression/correlation/matrix diagrams, mistake proofing, failure mode and effects analysis (FMEA), process capability analysis (PCA), design of experiments (DOE), matrix data analysis, etc.	Problem solving, checking, data collection and presentation, structuring ideas, identifying relationships, identifying control parameters, monitoring and maintaining control, etc.

			Reduction of inventory
Lean manufacturing	JIT	One-piece flow, pull system, takt time, leveled production, cell manufacturing, visual control, kanban, etc.	
	Total Productive Maintenance (TPM)	Overall equipment effectiveness (OEE), single-minute exchange of die (SMED), 5S, autonomous maintenance, planned maintenance, quality maintenance, etc.	Reduction of machine changeovers and breakdowns
	Autonomation	Mistake proofing, andon, full work system, etc.	Reduction of quality defects
	VSM	Current state map, future state map, flow diagrams, etc.	Identification of key processes and waste
	Kaizen events	5S, brainstorming, continuous flow, kanbans, data check sheet, five whys, Pareto chart, run chart, Gantt chart, VSM, process map, mistake proofing, etc.	Support and sustainment of Lean improvement initiatives
Lean Six Sigma	Same tier 2 methods and tools as for Six Sigma and Lean manufacturing		Reduction of waste and process variation

Table 6.1 Most Popular and Well-Known Business and Quality Improvement Models

Models	Tier 1 Methods	Tier 2 Methods	Tools	Objective/Application
	Agile manufacturing	Virtual enterprise (VE)	Virtual design, virtual manufacturing, virtual assembly, Internet-assisted manufacturing system, etc.	Quick response to market demands
		Physically distributed enterprise	Communication networks, e-mail, graphical user interface, etc.	Quick creation of alliances
		Rapid partnership formation	Multimedia, Internet, database, Microsoft project, case tools and electronic data interchange, QFD, benchmarking, etc.	Quick development of cooperative support
		Concurrent engineering	QFD, functional analysis, computer-aided manufacturing (CAM), finite element analysis, etc.	Shorter product development cycles
		Integrated product/ production/ business information system (BIS)	Multimedia, Internet, communication network, etc.	Achieve configurability
		Rapid prototyping	Virtual prototyping, CAD/CAM, etc.	Reduction of time to develop a product
		Electronic commerce	Electronic data interchange, etc.	Promotion of commerce

Quick response manufacturing (QRM)	The power of time	Manufacturing critical path time, etc.	Awareness and measure of lead time
	Organizational restructure	Cellular structure, team ownership, cross-trained workforce, lead time reduction goals, etc.	Creation of a lead time reduction-oriented organizational structure
	System dynamic	Capacity planning, etc.	Plan for spare capacity
	Unified strategy	Cell structure for office, simplified MRP, local suppliers, lead time reduction for suppliers, new product development lead time, paired-cell overlapping loops of cards with authorization (POLCA), etc.	Creation of a unified oriented organizational strategy
Theory of constraints (TOC)	Five focusing steps of TOC	System's constraint identification, exploitation of constraints, subordination to constraints, elevation of system's constraints, repetition of cycle	Improvement of organization's constraints
	Thinking process	Current reality tree, conflict resolution diagram, failure reality tree, prerequisite tree, transition tree	Create solutions to policy constraints
Business process reengineering (BPR)	Principles	Global and process orientation, value focus, concurrency, modularity, virtual resources, etc.	Guide the BPR initiative
	Process	Mission and vision, improvement opportunities identification, maintenance of systems, trade-off analyses, etc.	Set environment for BPR initiative
	Methods and tools	Integration definition (IDEF), IDEF 1, IDEF 3, IDEF 5, simulation, etc.	Operationalization of BPR principles

The objective of this section is simply to provide a categorization of business and quality improvement methods and tools with the aim of making their selection easier and more effective. Guidance for those who wish to extend their knowledge on the use and implementation of particular models, methods, or tools is provided in the references and further suggested reading sections at the end of this chapter.

6.3 Selection Criteria

In selecting the most appropriate business and quality improvement models, methods, and tools, various selection criteria should be considered. A selection criterion would provide a decision parameter for an organization to evaluate whether the implementation of a specific business or quality improvement approach is not only necessary but also possible. We consider four key organizational factors as part of the selection criteria: needs, cost–benefit, resources, and capabilities.

6.3.1 Criterion 1—Needs

Organizational needs in terms of the adoption of business and quality improvement initiatives are absolutely vital; they must be met if an organization is to prosper. The consequences of failing to meet these needs and thus adopt the right business and quality improvement models, methods, and tools are far reaching. This can result in low productivity, customer dissatisfaction, declining profit, and low morale among the workforce, in addition to other negative effects. However, for an organization to meet its business and quality improvement needs, it must first identify all of them. The QMS diagnostic methodology proposed in Chapter 4 will help an organization not only evaluate the maturity of its QMS and the effectiveness of its business processes, but also identify some of its primary needs in these areas. In particular, the QMS diagnostic methodology will help an organization answer some general questions, such as

Maturity diagnostic instrument (MDI):
- What is the attitude of the organization toward business improvement and quality?
- Are business improvement and quality initiatives sustained and aligned to the business plan and strategy?

- Does the organization apply a selection of business and quality improvement models, methods, and tools?
- Are top management and staff committed to business improvement and quality?

Self-assessment:

- What are the opportunities for improvement in the organization's core business processes?

Quality audits:

- Does the organization meet the quality standards required by its customers, suppliers, partners, collaborators, industry sector, or government regulations?

Answering the above questions will uncover the organization's business improvement and quality needs. For instance, if the MDI indicates that top management or staff are not committed to improvement initiatives, then it would be necessary for the organization to adopt or develop a strategy to achieve such engagement. Clearly, in this example, achieving top management or staff engagement in improvement activities is an organizational need that has to be met.

6.3.2 Criterion 2—Cost–Benefit

Organizations vary widely in nature and size and in the type of goods and services they produce or provide. They can be public or private, or profit or nonprofit enterprises. However, independently of their different characteristics, all organizations, even public sector or nonprofit ones, are required to effectively and efficiently manage their financial resources in order to survive. It is therefore crucial to consider the payoffs, in financial terms, and costs that an organization may obtain and incur on, during, and after the implementation of a specific business or quality improvement model, method, or tool. Even when the diagnostic indicates that an organization has the need to adopt a specific model, method, or tool, if the implementation and management costs exceed the financial benefit, then its implementation is doomed to fail in the long term. No organization will leave a business or quality improvement initiative running for long if instead of benefiting the company, it is causing it to lose money. It is for this reason that we consider cost–benefit as an important criterion to be included in selecting the right business and quality improvement models, methods, and tools. A cost–benefit analysis can provide a practical way of assessing, from a financial point of view, the desirability of the implementation of any improvement model, method, or tool. It can also justify the prioritization and selection of these by demonstrating their financial benefits in relation to their cost. In this context, those models, methods, and tools considered to be capable of meeting the needs highlighted by the QMS diagnostic can be compared, and those which more marginally outweigh their costs can be selected for implementation.

6.3.2.1 Difficulties with Cost–Benefit Analyses

A fundamental problem with cost–benefit analyses is that in most of the cases it is easier and more accurate to estimate the costs than the benefits. Costs come from claims on resources, such as the amount of staff time and training required to carry out the implementation of an improvement initiative, the purchase of physical resources, etc. In contrast, benefits are mere predictions of future events that may or may not occur. In addition, it can be difficult to calculate intangible benefits such as staff motivation and job satisfaction, improvement in the work environment, improvement in customer and supplier relations, etc. We therefore recommend getting the financial department to conduct a cost–benefit analysis with the support of the people and experts involved in the implementation of the business or quality improvement initiatives. This will also give more credibility to the cost–benefit analysis, as it would be performed by experts from the financial and accounting departments.

6.3.3 Criterion 3—Resources

All organizations bring together different resources in order to achieve their goals. These resources are the fuel that organizations need to keep going and produce the goods and services they provide to society. This is also the case for improvement initiatives, which require organizations to bring together and put in place certain resources for their effective implementation, functioning, and sustainment. The effective implementation, functioning, and sustainment of improvement initiatives cannot be achieved unless an organization has, or acquires, the appropriate resources. If an organization cannot supply such resources, it will not be able to support its improvement activities or ensure their implementation or sustainment. It is therefore essential, as part of the selection criteria, to consider whether an organization has, or needs to acquire, specific resources when selecting the right business or quality improvement models, methods, and tools. Resources refer to basic human, physical, financial, or information inputs that are not productive in themselves (unless converted into capabilities) but that can be called on when necessary. In particular, resources needed to support QMSs, business processes, and improvement initiatives include space, tools, money, machines, equipment, materials, personnel, plant facilities, software and hardware, and all other assets that may contribute to their implementation, functioning, and sustainment.

6.3.3.1 Identification of Resource Needs

Identification of resources means determining resource needs. The self-assessment process, using the selected BEM during the QMS and business processes diagnostic, will help an organization evaluate its effectiveness in terms of how it manages, utilizes, and preserves its current resources. As reviewed in Chapter 2, BEMs such as EFQM and Malcolm Baldrige address this through some of their evaluation criteria. Although this will provide an organization with an opportunity to understand its weaknesses in relation to its resources, the self-assessment process will not specifically indicate what resources the organization needs in order to implement the right business and quality improvement approaches. This has to be determined by top management by first defining what specific human, financial, physical, and information resources are needed, and then defining whether the organization currently has them. Depending on the needs of the organization as indicated by the QMS and business processes diagnostic, the resources needed to implement a specific model, method, or tool can widely vary across organizations. The organization's maturity level also plays an important role in this. For example, more mature organizations will certainly have more business and quality improvement-oriented resources already in place, which may not be the case for organizations falling within the classification of "uncommitted," "drifters," or "tool pushers."

6.3.3.2 Allocation of Resources

Once the resources needed to implement and maintain a high level of performance of the selected models, methods, and tools have been identified, the provision of these is the responsibility of top management. For instance, top management must provide an adequate number of personnel (e.g., human resources) that are qualified in terms of having the appropriate education, training, or experience to implement and manage the selected approaches.

Top management must also ensure that the organization provides and maintains the physical resources needed to conduct the necessary operations related to such models, methods, and tools. Specific physical resources could include, for example, infrastructure such as facilities and space to meet, testing and calibration equipment and laboratories, computerized systems, and any other physical resource required in supporting the improvement activities. The allocation of financial resources for the implementation, effective functioning, management, and maintenance of the models, methods, and tools selected also falls within the responsibility of top management. When discussing selection criterion 2, we highlighted the fact that the adoption of any improvement approach would incur a cost to the organization. However, we also emphasized the importance of determining whether the expected financial benefit would outweigh the investment and sustainment costs and only recommended implementing the model, method, or tool if this was the case. In any case, top management must take responsibility for observing that appropriate and sufficient financial resources are allocated to cover the initial investment and subsequent costs of sustainment. Finally, top management must also make sure that information resources, such as technical data and information in all forms, are available and accessible as required by the personnel implementing the models, methods, and tools.

6.3.4 Criterion 4—Capabilities

Similar to a lack of resources, a lack of certain specific organizational capabilities can also hinder the successful implementation, functioning, and sustainment of improvement activities. It is for this reason that when selecting the right models, methods, and tools, certain organizational capabilities must be considered as part of the selection criteria. In order to clearly understand this criterion, it is first necessary to differentiate capabilities from resources. Although the terms *resources* and *capabilities* are sometimes used interchangeably, they are technically not the same. Resources are not productive inputs of a singular nature, whereas capabilities are best referred to as the integration of various resources in a way in which boosts an organization's competitive advantage. For example, statistical process control (SPC) software designed to analyze data from a production line can be considered a resource. However, this software may be of no value until it is integrated into a QMS that clearly indicates to the organization what to do with the analysis provided by the software and oversees the implementation of any corrective or improvement action derived from such analysis. Some authors also differentiate organizational capabilities from individual competences. However, since individual competences may also procure a strategic advantage when effectively integrated into the organization's improvement activities, we also refer to them as capabilities.

Based on the literature and experience, we consider the following to be some of the essential capabilities that organizations must have or develop in order to implement, manage, and sustain any business and quality improvement models, methods, and tools:

- Top management commitment and involvement in continuous improvement (CI) activities (MDI)
- Staff commitment and involvement in CI activities (MDI)
- An organizational culture that supports and aids change (MDI)
- Effective internal communication among different hierarchical levels and staff as well as external communication with suppliers, customers, and third parties
- Strong leadership traits capable of exhibiting excellent project management styles
- An organizational environment that encourages teamwork, trust, friendship, and positive informal relations among groups
- Ability to share knowledge
- An organizational culture that supports the continuous education and training of managers and people directly involved in the company's improvement activities (MDI)
- An organizational culture that is customer focused and recognizes the importance of CI (MDI)
- Ability on the part of top management to link the selected models, methods, and tools with the organization's strategy and planning (MDI)
- Understanding of and expertise on the selected models, methods, and tools (MDI)

6.3.4.1 Identification of Capability Needs

The Dale and Lascelles' (1997) maturity classification evaluates organizations based on the degree of development of certain organizational capabilities that support TQM. Because it was developed by us in reference to this model, the MDI can help organizations to determine whether they have the necessary capabilities to implement, manage, and sustain a business and quality improvement model, method, and tool. For example, some of the subcategories of the MDI (e.g., 5, 11, 13, 16, 21) evaluate the degree of engagement of top management and staff in quality and CI improvement activities. Scores above 4 in those subcategories would denote a weak commitment of top management and staff. The closer the score is to 7, the weaker the commitment is. Obviously, this would indicate that the organization does not, at that moment, have the capability of effectively implementing, managing, or sustaining the selected business or quality improvement models, methods, or tools.

Capabilities that the MDI can shed some light on as to whether the organization possesses them are indicated in the above bullet point list as *(MDI)*. However, the organization will need to find a way of evaluating those capabilities for which the MDI does not provide information. This can be done by cross-referencing the capability needed with already-used measures of performance. An example of this is the capability that refers to "an organizational environment that encourages teamwork, trust, friendship, and positive informal relations among groups." According to the management and organizational behavior theory, symptoms such as high turnover and absenteeism can indicate whether an organization's environment is conducive to teamwork, trust, and friendship, or whether it promotes positive, informal relations among groups. If these measures are not employed by an organization, an alternative approach could be to conduct a survey investigation with the organization's staff to find out whether that environment exists.

Similar to the development of resources, the development of the organizational capabilities needed to implement, manage, and sustain any business or quality improvement approach falls within the responsibility of top management.

6.4 Selecting the Right Models, Methods, and Tools

In this section we present a methodology for guiding organizations in the selection of the right models, methods, and tools once they have understood the status of its QMS and business processes. Since the diagnostic of the QMS and business processes as well as the alignment of the improvement plan with the organization's strategic planning are previous stages to this selection methodology, we start this section with a brief review of these stages.

6.4.1 Previous Stages to the Selection of Business and Quality Improvement Models, Methods, and Tools

In Chapter 4 we introduced a diagnostic methodology intended to help an organization do a thorough evaluation of the current status of its QMS and business processes. In particular, and as illustrated in Figure 4.3, this methodology consists of three main evaluations that include: (1) a maturity diagnostic, (2) a self-assessment process, and (3) quality management audits. Specifically, this diagnostic methodology can help an organization to

Maturity diagnostic:
- Define the current maturity level of its QMS
- Set a before and after improvements comparative platform
- Identify the specific strengths and limitations of its QMS and improvement activities and thus determine business and quality improvement needs
- Determine whether an organization possesses some of the organizational capabilities needed to successfully adopt, manage, and sustain those business and quality improvement models, methods, and tools identified as the ones that can enhance its QMS and business processes

Self-assessment process:
- Identify strengths and opportunities for improvement in the organization's business processes

Quality audits:
- Determine whether the organization's processes comply with the quality standards required by its customers, suppliers, partners, collaborators, industry sector, or government regulations

In summary, the diagnostic methodology will provide an organization with a clear understanding of its current QMS and business processes. Once this has been understood, the organization can then propose and deploy an action plan to address the areas for improvement highlighted in the overall diagnostic of its QMS and business processes. At the end of Chapter 4 we discussed the importance of aligning the improvement plan to the organization's strategy and planning, while in Chapter 5 we provided a method for achieving this. Selecting the appropriate business and quality improvement models, methods, and tools to fulfill the organization's improvement plan is considered to be the next stage.

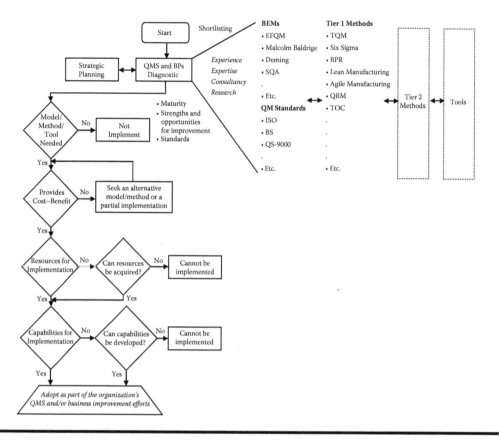

Figure 6.1   Illustration of the selection methodology.

6.4.2 Selecting the Right Models, Methods, and Tools

Figure 6.1 illustrates a methodology that can be used to guide an organization in the selection of the right models, methods, and tools needed to enhance its QMS and business processes. In general terms, the selection methodology indicates that six steps have to be carried out in order to conduct the selection. These steps include the following:

Step 1: QMS and business processes diagnostic—alignment of action plan with an organization's strategic planning. The selection of the right models, methods, and tools should start with the identification and understanding of the areas for improvement in the organization's QMS and business processes, and then by formulating and aligning an improvement plan to its strategic planning. Chapters 4 and 5 have been dedicated to helping an organization carry out this first step, while Section 6.4.2 provides a brief summary of these activities.

Step 2: Models, methods, and tools shortlisting. Once the QMS and business processes diagnostic has been completed, its results will provide the organization with a clear picture of the organization's actual situation. Based on this, the organization will have to "shortlist" some of the business or quality improvement models and methods it believes can help it overcome the weak areas indicated by the diagnostic. As previously mentioned, a huge number—probably hundreds—of models, methods, and tools have been developed to help organizations improve their operations and the quality of their processes, products, and services. As a general and nonexhaustive guide, the organization can use Table 6.1, from which it can select a small group of models or methods that it thinks can be used to reduce or overcome the weaknesses highlighted by the diagnostic. To facilitate the selection, we recommend that the shortlist focus on models and tier 1 methods only. Once the selection of tier 1 methods has been done, the tier 2 methods and tools attached to it can also be evaluated to find out their suitability in helping the organization with the problems highlighted by the QMS and business processes diagnostic.

Some organizations, especially large ones, may certainly have knowledge, expertise, or experience concerning the use of some of the models and methods included in Table 6.1. These organizations can use such know-how to more effectively shortlist the most adequate models or methods to tackle its weaknesses. For example, if the self-assessment process indicates that "critical to success processes have not been clearly identified," then the organization's knowledge, expertise, or experience can be used to indicate that Lean manufacturing should be shortlisted, as value stream mapping (VSM) may help it with this identification. However, SMEs may not have this knowledge, expertise, or experience in-house; in this case, the organization will have to research or consult with experts about what models or methods may help in addressing its problems. Consultancy does not necessarily entail paying a fortune to an expert to get some guidance or information. Nowadays, informal consultancy, or better named knowledge sharing, can be obtained by receiving some informal guidance from colleagues, for example, from professional institutions as well as from suppliers, customers, or even local universities. Publications from professional institutions and related textbooks can also be used to provide a more clear understanding of the approaches included in Table 6.1. This will enable the initial selection of the models and methods that may help address the organization's problems. As previously mentioned, guidance for those who wish to extend their knowledge on particular models and methods is provided in the references and further suggested reading sections at the end of this chapter.

Step 3: Evaluating the need of implementing the shortlisted models, methods, and tools. Once the models and methods have been shortlisted, the next step in the selection methodology is to evaluate, in more detail, whether these are really needed by the organization. In other words, this is done to evaluate whether such models or methods are the most adequate for overcoming, or at least reducing, the problems highlighted by the QMS and business process diagnostic. Here, it is very important to understand what the issues highlighted by the diagnostic are, as well as what the general objectives of the models and methods are so that they can be matched. For instance, if the quality management audit carried out as part of the diagnostic indicates that a process does not comply with customer requirements, an improvement of such a process may be needed. As an initial step, the organization may need to discover the root cause of the problem in order to understand the noncompliance and later tackle it. This problem can be matched, for example, with Six Sigma, as it offers DMAIC and tools such as cause-and-effect analysis, which would aid in uncovering the root cause of the problem and facilitate its elimination.

If the evaluation indicates that the problem and objective of the shortlisted models or methods do not match, then this demonstrates that the organization does not need to implement a particular model or method. In this case, and as illustrated in Figure 6.1, the shortlisted model or method should not be implemented. Here, we also recommend that the organization do an evaluation for every tier 2 method and tool associated with a shortlisted tier 1 method. For example, if Six Sigma has been identified and evaluated as a possible tier 1 method for tackling a weakness, then all of its tier 2 methods and tools should also be evaluated. In many cases, not all tier 2 methods and tools will need to be implemented or used to solve a specific problem.

Step 4: Evaluating the cost–benefit of implementing the shortlisted models, methods, and tools. If in step 3 the needs evaluation results indicate that the objective of the shortlisted models or methods matches the problem, and thus are required to be implemented, the next step is to evaluate their cost–benefit. As previously discussed, it is imperative that the organization obtain some financial benefit from the implementation or use of the selected models or methods. The implementation, management, and sustainment of the shortlisted models or methods will require a financial investment, which by no means should be higher than the expected financial benefit. If this is the case, our selection methodology indicates that such models or methods should not be implemented, but that alternative ones or a partial implementation of them should be sought. For example, in the last example, where the process did not comply with customer requirements, the organization may not need to implement Six Sigma on a large organizational scale, which would obviously require a huge financial investment. Instead, it may decide to use Six Sigma, DMAIC, and cause-and-effect analysis as a one-off approach to specifically solving this problem. This can largely reduce the cost of solving the problem. Many authors and experts may refute a partial implementation of a model or method, as they may consider it a "quick fix" or short-term approach to business improvement. Although we agree with this fact, we also believe that this approach, in some cases, may help organizations with limited financial resources to support a full implementation. Some general information about cost–benefit analyses is included in Section 6.3.2.

Step 5: Evaluating whether the organization possesses the required resources to effectively implement, manage, and sustain the short-listed models, methods, and tools. As previously stated, discussing the resources criteria, implementation, management, and sustainment of business and quality improvement models and methods will consume human, physical, financial, and information resources. An organization may need to implement or use specific models, methods, and tools, which can also be determined to bring a financial gain to the company. However, if the organization does not have the necessary resources, it will not be able to implement them. For this reason, the selection methodology illustrated in Figure 6.1 indicates that the organization must evaluate whether it has the resources needed. If it does, then the organization can move on to the last selection step, but if it does not, then top management will have to make sure that the necessary resources are acquired. Otherwise, the implementation of the models, methods, and tools will not be possible. We provided some general guidance about resource identification and allocation in Section 6.3.3.

Step 6: Evaluating whether the organization possesses the required capabilities to effectively implement, manage, and sustain the shortlisted models, methods, and tools. As is the case with resources, an organization must have certain internal capabilities in order to effectively implement, manage, and sustain the shortlisted business or quality improvement models, methods, and tools. It is for this reason that a capability assessment has also been included as part of the selection methodology. An organization may have determined that it requires the implementation of a specific model, method, or tool, as well as the fact that its implementation or use will bring about a financial benefit. Moreover, the organization may have also determined that it has the resources needed to implement, manage, and sustain the selected approach. However, if it does not integrate such resources in a way that will create a competitive advantage, in other words, if it does not transform them into capabilities, then the implementation, operation, or sustainment is destined to fail. In Section 6.3.4 we provide a list of the capabilities needed to effectively implement, manage, and sustain business and quality improvement approaches, as well as how to identify whether an organization possesses them.

6.4.3 Diagnosis and Selection of the Right Models, Methods, and Tools

If after the evaluations of the shortlisted models, methods, and tools have been determined to

1. Be needed by the organization, as they may solve the company's weaknesses highlighted by the diagnostic
2. Provide a cost–benefit, as their cost of implementation, management, and sustainment will not exceed the financial benefit that the organization may obtain from them
3. Be capable of being effectively implemented, managed, and sustained because the organization has, or can acquire, the resources needed
4. Be capable of being effectively implemented, managed, and sustained because the organization has, or can develop, the capabilities needed

then these models, methods, and tools can be adopted as part of the organization's QMS or business improvement efforts.

Here, it is essential to mention that the adopted models, methods, and tools will only address, in most cases, operational rather than management and organizational behavior problems. Due to the broad scope of evaluation yielded by the QMS and business process diagnostic, many problems that are not related to operations may be highlighted. For example, the QMS maturity evaluation may uncover that top management and staff do not get involved in CI activities. Similarly, the self-assessment process may also reveal problems with the organization's leadership, formulation of policies and strategies, management of resources, etc. If this is the case, then the implementation of a business and quality improvement model or method will have only a slight effect, if any at all, on such problems. The implementation and use of the approaches included in Table 6.1 will certainly require a cultural change if a business is to be managed according to these philosophies and principles. However, we consider that the problems given in the previously mentioned examples should be addressed by means other than by implementing a business or quality improvement approach. In such scenarios, the organization would need to seek appropriate actions to tackle these problems.

Some of the areas that may be highlighted as problematic by the QMS and business processes diagnostic, and that we consider cannot be improved by implementing or using a business or quality improvement approach, include (1) lack of top management and staff support and involvement in CI activities, (2) difficulties in the sustainment of business and quality improvement approaches and activities, (3) inability to effectively manage change, (4) organizational culture that does not support CI initiatives, (5) lack of integration of CI activities with the organization's strategy, (6) lack of a customer-focused culture, (7) organization does not place a positive value on internal and external relationships (e.g., with customers, suppliers, employees), (8) leadership, (9) people management, and (10) resources management, among others.

6.5 Summary

In this chapter we have focused on the selection of the right models, methods, and tools that need to be adopted as part of an organization's QMS or business improvement efforts to meet the improvement opportunities identified by the diagnostic. To facilitate the selection, we classified, based on their purpose and characteristics, the most popular and well-known business and quality improvement approaches into models, tier 1 and tier 2 methods, and tools.

We also defined a selection criterion that we consider needs to be taken into consideration in assessing the suitability of an improvement approach to help an organization overcome weaknesses and meet its improvement needs. The selection criteria evaluate not only whether the selected improvement approaches are needed, but also whether they will provide a financial benefit; moreover, they help determine whether the organization has the required resources and capabilities for their implementation, management, and sustainment.

Finally, in this chapter we defined a series of steps that an organization can follow in order to select the most appropriate models, methods, and tools. These steps consists of (1) performing the QMS and business processes diagnostic and understanding its results; (2) shortlisting a small group of models or methods, specifically those whose objectives match the company's improvement needs; (3) evaluating whether an organization really needs the shortlisted approaches by matching the objectives of the approaches with the organization's improvement needs; (4) evaluating whether the improvement approaches may provide financial benefits; and (5) assessing whether the organization has the resources and (6) capabilities needed for their effective implementation, management, and sustainment. These are considered crucial steps to the effective selection of the right improvement approaches. In Chapter 7 we look at how such models, methods, and tools can be effectively implemented.

6.5.1 Key Points to Remember

- Selecting the right models, methods, and tools is essential to successfully design, implement, or improve an organization's QMS and business processes.
- To assist in the selection of the most appropriate models, methods, and tools we have provided in this chapter a nonexhaustive classification and alignment of some of the most popular and well-known business and quality improvement approaches currently used by organizations.
- In selecting the most appropriate models, methods, and tools we consider needs, cost–benefit, resources, and capabilities as the four key organizational elements of the selection criteria.
- In this chapter we provide a methodology for guiding organizations in the selection of the right models, methods, and tools after they have understood the status of the QMS and business processes.

References

Beecroft, D. G. (2004). Evolving quality improvement/implementation strategies. *Annual Quality Congress Proceedings*, Vol. 58, No. 0, pp. 435–428.

Dale, B. G., and Lascelles, D. M. (1997). Total quality management adoption: Revisiting the levels. *TQM Magazine,* Vol. 9, No. 6, pp. 418–428.

Hagemeyer, C., Gershenson, J. K., and Johnson, D. M. (2006). Classification and application of problem solving quality tools: A manufacturing case study. *TQM Magazine*, Vol. 18, No. 5, pp. 455–483.

Montgomery, D. C. (2009). *Statistical quality control: A modern introduction*, 6th ed. John Wiley & Sons, Hoboken, NJ.

Pyzdek, T. (2003). *The Six Sigma handbook: A complete guide for green belts, black belts and managers at all levels*. McGraw-Hill, New York.

Slack, N., Chambers, S., and Johnston, R. (2009). *Operations management*, 6th ed. FT/Prentice-Hall, London.

Further Suggested Reading

Cox III, J., and Schleier, J. (2010). *Theory of constraints handbook*. McGraw-Hill, New York.

Dale, B. G. (2003). *Managing quality*, 4th ed. Blackwell Publishing, Oxford.

Grover, V., and Kettinger, W. J. (1995). *Business process change: reengineering concepts, methods and technologies*. IGI Publishing, Hershey, PA.

Gunasekaran, A. (2001). *Agile manufacturing: The 21st century competitive strategy*. Elsevier Science Ltd., Oxford.

Oakland, J. S. (2003). *TQM*, 3rd ed. Butterworth-Heinemann, Oxford.

Suri, R. (1998). *Quick response manufacturing: A company wide approach to reducing lead times*. Productivity Press, Portland, OR.

Chapter 7

Chapter 7

QMS Implementation

7.1 Introduction

Previous chapters have elaborated on the quality management system (QMS) and its significance in the existing competitive scenario. We have attempted to explain different business models, quality management standards, models, methods, and tools that are available for an organization looking to design, implement, or improve a QMS. In Chapter 3 we further argued why organizations need to understand and visualize their processes. We have very clearly conveyed the message that the quality is very much dependent on the way processes are designed and delivered. A poorly designed and unreliable process will always generate errors and quality issues, no matter how hard those who make use of these processes try. We further emphasized the necessity of developing IT competence and the role of value stream mapping in identifying non-value-added activities. Chapter 4 focused on QMSs diagnostics, whereas Chapter 5 provided a brief overview on strategic quality planning. Finally, Chapter 6 emphasized the importance of selecting the right methods and tools. In Chapter 6 a methodology was also proposed to guide an organization in the selection of the right models, methods, and tools that are needed for a QMS and business processes improvement.

In this chapter our primary focus is to discuss the QMS implementation process. We will illustrate how the proposed selection methodology explained in Chapter 6 can be implemented. The chapter provides a brief overview of the challenges that management needs to overcome during the QMS implementation process. We then discuss some of the major critical success factors (CSFs) for QMS and emphasize the need for an awareness of certain barriers. Finally, we detail the challenges in managing change within an organization and highlight how the proposed methodology links up with the CSFs. In summary, this chapter provides practitioners with the essential requirements for successful QMS implementation.

7.2 QMS Implementation Challenges

We have made it very clear in previous chapters that QMS is the most pressing need of the current era, and organizations cannot afford to ignore quality-related issues. So far we have also looked at the various aspects of managing quality by discussing different quality models, methods, and tools, defining and improving processes or strategic quality planning. The actual implementation of QMS is, however, a major issue for organizations. Regardless of how well the QMS is planned, it does not deliver any value unless it is well implemented within the organization. The significance of implementation is also important from an organization's strategy viewpoint. In the strategy-making process organizations also have to make sure that whatever strategy they are going to adapt is well executed, as their organizational performance hinges on how well their strategy is executed. Therefore, we would like to emphasize that good QMS implementation is key for better organizational performance.

With the understanding that proper execution or implementation is a much needed requirement for the intended benefit of QMS implementation, organizations often struggle at this stage. A thorough understanding of the key factors that influence the QMS implementation is necessary. There are several factors that pose substantial challenges to the management of an organization. For example, organizations need to have an adept and decisive leadership who can make instant and effective decisions, as failure to do so can completely jeopardize the QMS implementation. Empowering employees, improving processes, instituting a quality-oriented culture, and promoting teamwork ethics are also some of the other challenges that an organization has to overcome. Organizations trying to implement a quality improvement framework continuously seek to identify factors that are believed to be critical to successful implementation and are often termed critical success factors (CSFs). There are a number of CSFs that, when aligned, will result in a successful QMS implementation in an organization. Organizations failing to understand and minimize/eliminate these CSFs may struggle to implement QMS and fall short of their goal of enhancing their performance. Keeping in mind the major role of the CSFs, the next section elaborates these CSFs in detail.

7.3 Critical Success Factors for the Implementation of the Selection Methodology

The previous section highlighted the fact that QMS implementation is not a layman's job, and it needs a thorough investigation of all the underlying factors that are critical for its successful implementation. CSFs are often referred to as actions and processes that can be controlled by management to achieve the organization's goals. Organizations must ensure that CSFs are achieved since they are a source of competitive leverage. In this section we attempt to identify these CSFs based on the selection methodology we presented in Chapter 6. But before we discuss the different critical factors in detail, let's revisit the core focus of QMS. QMS focuses on understanding, controlling, and improving work processes. The goals of QMS are also to analyze the causes of variability, take suitable steps to make the work process predictable, and seek continuous process performance improvement. To achieve these goals, the management of an organization must build cross-functional teams to identify and resolve quality problems. This would also involve analyzing and monitoring the processes using tools such as process mapping or flowcharts and collecting useful information that can explain the nature of a problem, so that the necessary improvement steps can be taken.

In the previous chapter we suggested four steps for the selection methodology where management first needs to evaluate all the shortlisted models, methods, and tools, since this evaluation will assist organizations in choosing the right tool that can complement their strength. Understanding of management's commitment and identification of process improvement opportunities is vital for the successful QMS implementation. Second, management needs to perform the cost–benefit analysis to judge the financial viability and intended financial benefits of the QMS implementation. If management finds that the cost of QMS implementation exceeds the intended benefits, then it is an early warning sign indication that the implementation may fail in the long run. Another important aspect of QMS implementation is to evaluate whether an organization has enough of the resources that are required to implement the shortlisted models. If organizations do not possess the required resources, they need to acquire those that specifically support QMS, business processes, and improvement initiatives. Finally, management needs to evaluate

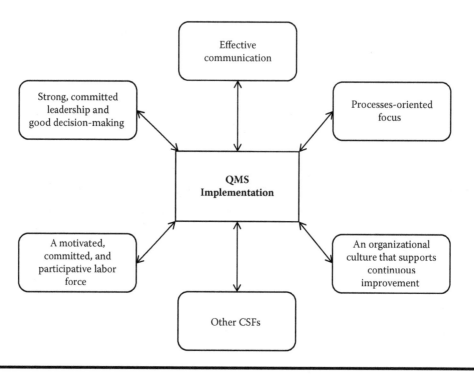

Figure 7.1   Critical success factors (CSFs) for QMS implementation.

whether an organization has the capabilities that are needed to implement the shortlisted models, such as top management and staff commitment and involvement in continuous improvement (CI), organizational culture, strong leadership, and effective internal communication. The lack of these specific organizational capabilities can hinder the successful implementation, functioning, and sustainment of improvement activities. Therefore, it is now clear that there are a number of factors that are crucial for the successful implementation of QMSs. Although there are many critical success factors, as suggested by a number of practitioners and researchers (Coronado and Antony, 2002) for continuous improvement, in the upcoming subsections our focus is to give you a brief idea on some of the major CSFs (Figure 7.1) necessary for the successful implementation of QMSs.

7.3.1 CSF 1: Strong Committed Leadership and Good Decision Making

As we have already discussed, there are several critical success factors that hold the key for a successful QMS implementation. Among those, we identify strong and committed leadership as one of the major CSFs. Although quality is a key issue among manufacturing organizations, and its importance is well acknowledged, the implementers of quality improvement initiatives often fail to recognize the importance of people in the process. They fail to understand how the social interaction of the different people involved in the process with each other increases the productivity and profitability of an organization. One of the significant roles is played by leaders who drive organizational success by motivating and engaging people involved at different levels within an organization. Examples from the business world clearly recognize the role of leadership in driving and reviving organizations from the brink of collapse to a highly successful ladder. For example, Apple Inc.'s success story is not hidden from anyone, where one can see how their former visionary leader Steve Jobs revived the company from a declining phase in the late 1990s to transform the organization into one of the most highly valued companies in the world, with market capitalization of more than $600 billion in the year 2012. The majority of the business world would credit this success of Apple to a great extent to Steve Jobs' charismatic and visionary leadership ability. The Apple story is just one of the many that highlights the role of leadership in an organization's success. The importance of good leadership in creating what is required of an organization is accepted indisputably, from small teams to global enterprises. There are numerous success stories involving organizations with charismatic leaders, such as Sir Richard Branson of Virgin Group, Michael O'Leary of Ryanair, or Henry Ford of Ford Motors. All these examples show that a strong and committed leadership is essential to an organization's success.

A flexible and adaptable leadership is critical to any group environment, and it exists at all levels throughout an organization. Research studies (Wang et al., 2005; Thite, 2000) have highlighted that essential leadership traits and abilities, such as the ability to manage people, stress, emotions, bureaucracy, and communication, are required to ensure success. Charismatic leadership behaviors are identified as among the most critical leadership behaviors in terms of satisfaction. Charismatic leaders attempt to fuse each member's personal goals with the organizational mission that promotes team commitment and cohesiveness leading to improved performance. The world has seen many charismatic leaders in the last century who have made a big impact on the success map. An important trait of leadership is also to be visionary, and he or she must also have the ability to create and support an empowering atmosphere that assists self-directed teams in adapting to environmental changes. Leaders have a complex task of producing and managing periods of stability as well as developing vision and planning future strategy. And to execute these tasks successfully, leaders have to use their expertise, commitment, decisiveness, quality, vision, and charismatic personality to develop team competence, commitment, group expertise, and group support.

So if we look at the leadership trait from a QMS implementation perspective, it is obvious that leaders set a direction and a standard of excellence for an organization. And to successfully implement a QMS within an organization, their role and contribution is significant. An important question at this stage is: How can organizations develop such leadership characteristics to complement QMS implementation? One can argue that those qualities of excellent leadership cannot be developed, as leaders are born, not made. But we do not want to get into the debate here about whether leaders are born or made. Instead, what we wish to argue is that organizations must choose a leader who possesses good leadership qualities because it is important to note that unless an organization has a leader who possesses all the good leadership qualities, it becomes a cumbersome task to execute any quality improvement initiatives. A good leader with strong commitment, work ethics, and good decision-making skills is required to achieve success in the design, implementation, or improvement of QMSs. Making the right decision as per the need of the time is equally important as guiding the organization in the right direction. Another trait of successful leadership also involves having good decision-making skills, and often the right decision made in the early stages helps an organization to cope with the difficulties that arise at later stages. Failure to make the right decisions can also act as a major barrier to an organization's growth and quality improvement plans. Organizations therefore need to choose leaders who either possess those qualities or show the acumen to develop them over time if they want to climb the success ladder. Referring to our four steps of the proposed selection methodology, we would like to emphasize that without a strong and committed leadership, it is quite hard to follow these steps and achieve the desired benefits. Therefore, we would like to conclude that strong, committed leadership and good decision-making skills are vital for any organization that wishes to implement QMS successfully.

7.3.2 CSF 2: Motivated, Committed, and Participative Labor Force

Though leadership commitment is a key to QMS implementation, it must not be forgotten that the success in any quality improvement initiative is only possible through the commitment of teamwork. Hence, unless the whole organization works as a team, it will be almost impossible to implement any quality improvement initiatives. Therefore, a same level of motivation, commitment, and participation is required from the top management as well as employees at all levels of an organization. This motivation and commitment has to be driven from the leaders and top management through their active involvement in the QMS implementation process. Leading the implementation process gives a very strong message to their employees. A good example for all employees could be set by the top management showing their enthusiasm and strong commitment in driving the implementation of QMS. While doing so, leaders and top management must also ensure that QMS is well aligned with the strategic aims and objectives of the organization.

A lack of motivation and commitment can act as a hindrance to QMS implementation. Leaders and top management have to adapt a number of initiatives to build a supportive culture within the organization. They must ensure that their employees are well informed about all of the decisions and initiatives that management is currently pursuing. This can be done by the active involvement of the leaders and executive management team, and following certain hands-on approaches, such as conducting frequent quality improvement reviews (i.e., weekly or monthly), monitoring projects through weekly summary reports, and making site visits at manufacturing operations to ensure that QMS is being well integrated into the organization. In order to motivate the workforce even further, management needs to provide training activities. This will also ensure a smooth implementation, as chances of errors can be reduced significantly. Well-trained employees working in a quality supportive culture, under a strong and committed leadership, will certainly be of great assistance in the design, implementation, or improvement of QMSs. In addition, organizations can also develop a reward and recognition system to motivate their employees. Further, many organizations that have successfully implemented QMSs have performance appraisals, promotions, and recognitions linked to their implementation and success. Research evidence shows that the failure of many organizations to implement quality improvements plans is largely attributed to a shortfall in their employee motivation practices. Organizations must understand that recognition needs are vital not only for the leaders, but also for stakeholders and the team members. Recognition not only creates a "feel good" factor, but also leads to healthy competition. Therefore, an integrated employee reward and recognition framework must be adapted by organizations willing to implement a QMS. We would also like to emphasize that without the support and commitment of top management and leaders, it will be very tough to encourage their labor force to show their commitment, which is vital for the successful implementation of a QMS. Hence, organizations need to have both a strong and committed leadership and a highly motivated and committed workforce.

7.3.3 CSF 3: Process-Oriented Focus

Another key critical success factor for QMS implementation is an organization's process-oriented focus. We have argued the significance of process management and its intended benefits in Chapter 3. Understanding of processes is crucial since managing quality within the organizations is very much dependent on the way the organizations manage their processes. Most of the operational inefficiencies in organizations are attributed to poor process design and execution. Often organizations failing to understand their processes struggle hard to maintain their quality levels, which ultimately impacts on an organization's performance. On the other hand, organizations that have a deeply rooted process-oriented culture (i.e., teamwork, readiness to change, and customer focus) perform well. The notion behind an efficient process management is to improve the organization's work flow and make that organization capable of adjusting to the uncertain environment. This is possible when management is aware of which business processes are performed within the organization and how they are related to each other; thus, the design and documentation of process is an important element of a process-driven culture. IT systems also play a crucial role in process management since they complement business processes, and seamlessly support business processes, process-oriented organizational structure, people and expertise, and process-oriented HR systems. IT also integrates different business units through the end-to-end linking of value chains of one business unit with those of another business unit, thus supporting the interorganizational business processes.

Realizing the benefits that organizations perceive through a process-oriented culture, we would like to assert that in order to implement QMS successfully, management needs to understand, analyze, and continuously monitor their process improvement initiatives. Organizations must build a process-driven culture and develop unique IT capabilities. A strong and committed leadership can take the initiative in instituting process-oriented culture. We recommend managers use the value stream mapping tool to identify the value-added and non-value-added activities. After the identification of all non-value- and value-added activities, management can eliminate the non-value-added activities to make processes more efficient and concentrate on further improving the value-added activities. Management must also continuously monitor the performance of their processes, particularly their core processes, i.e., processes that are essential to the delivery of outputs and accomplishing business goals. Organizations implementing a QMS must ensure the consistency of core processes to respond quickly to the changing market conditions. In addition, management also needs to have a well-defined and well-assessed process improvement agenda that they must execute well. Thus, organizations following these suggestions do see immense benefits and are capable of successful QMS implementation. Therefore, a key to a successful QMS implementation is to establish a process-oriented culture within an organization.

7.3.4 CSF 4: Organizational Culture That Supports Continuous Improvement (CI)

Leadership and employee commitment is as vital as a process-oriented focus in organizations that are willing to implement QMS successfully. However, to ensure that organizations continue to follow the right path without any obstacles, an organizational culture that supports continuous improvement is essential. An organizational culture can be defined as "a system of shared values defining what is important, and norms, defining appropriate attitudes and behaviours, that guide members' attitudes and behaviours" (O'Reilly and Chatman, 1996). From the definition it is evident that an organization's culture comprises all of the values, beliefs, assumptions, principles, and norms that define how individuals and groups of people think, make decisions, and perform. Leaders are responsible for instituting an organizational culture, and most of the time it develops from the way the leaders behave in the organization. There are many examples around, and we have given a few examples of leaders who developed a unique culture based on innovation, such as Steve Jobs and Apple's innovation-oriented culture and Jack Welsh and the Six Sigma-integrated culture of GE. However, while creating the culture that supports quality improvement initiatives, leaders and their executive team will face strong resistance from employees unless people resisting cultural change understand the change first. An organization's cultural practices can be the biggest barrier, since they inhibit a quality improvement effort before it even starts. For example, if an organization operates by employee consensus, employees may find the top-down nature of continuous improvement disrespectful to their sensibility. Hence, in order to establish a successful CI-oriented organizational culture, top management needs to have a clear communication plan and channels, must motivate individuals to overcome resistance, and educate senior managers, employees, and customers on the benefits of QMS implementation.

Organizations are a blend of several different cultures and the subcultures that often develop over time, all of which contribute to the overall diversity found within the organization. In addition, there are several dimensions to an organizational culture that are closely linked to QMS values and beliefs, such as the basis of truth and rationality, motivation, stability vs. change, orientation to work, control, coordination, and responsibility. For instance, quality improvement initiatives follow an approach to truth and rationality through scientific method and data collection. This is an essential part of the QMS implementation process, as organizations looking to implement a QMS need to measure continuously their processes and look for ways of improvement. In addition, the understanding of various interrelations among the factors is complex and is only possible through the analysis of the collected data. Thus, organizational decision making must be based on the factual information; i.e., an organization must have a management culture that is driven by fact and not by experience or feelings. An organizational culture that focuses on motivating employees should make sure that systems are designed in a way that support their efforts, as often problems are caused by poor systems rather than the employees themselves. An organizational culture must also be developed in a way that employees share the same vision and goals as the organization and actively participate in the decision-making process. We would therefore like to assert that an organizational culture must be customer focused and should progress toward an internal process improvement, reduction of non-value-added activities, developing IT capabilities, and the identification of core processes. Moreover, the culture must promote cooperation and internal and external collaboration. Such a culture inheriting these attributes provides a supportive environment for QMS implementation. Executive management must also look at the organizational culture of other successful companies and attempt to adapt those good practices and build a culture that not only is aligned with the organizational aims and objectives, but also provides a supportive and friendly environment wherein employees work as a team to assist in successful QMS implementation. In summary, the key to a culture of continuous improvement is to be aware of the current culture, identify the elements that can be retained or discarded, design a culture for the future, share the vision, align leaders, empower and train employees, build the ethics of teamwork, involve everyone from top to bottom in decision making, celebrate achievements, treat culture as a strategic issue, remove cultural barriers, and keep the culture of the organization up to date.

7.3.5 CSF 5: Effective Communication

Effective communication is another critical factor for a successful QMS implementation. There is a strong relationship between good communication and successful quality implementation. Research evidence (Ocholi, 1998) has shown that quality management depends on communication that flows in all directions up, down, and laterally within an organization. In addition, if communication is used properly, it can act as an instrument to measure effective job performance, and serve as an index for employee motivation, leading to high productivity. In a corporate world effective communication is needed for mentoring and supervising. Leaders and executive management must follow effective means of communication with their employees, as it can help to build effective relationships at all levels, such as between supervisor, subordinates, clients, and even customers. Effective communication avoids the risk of misunderstanding and can easily change someone's wrong perception. This is particularly beneficial when leaders are attempting to create a CI-focused culture. Effective communication is much needed during tough times, such as during the process of organizational change when a lot of confusion arises at all levels. In such circumstances, only effective communication can resolve many issues, as it prepares the mind for change so that when anticipated changes take place, it helps to overcome the associated fear and panic. For example, if a need for the downsizing of a department arises and if this has not been communicated properly, employees will feel less motivated, and this will in turn affect the organizational performance, resulting in poor quality.

To establish effective communication management must ensure a two-sided channel of information flow based on common needs. The first step in this approach is to develop an ability to listen effectively to what employees have to say. Executive management must consult the managers and other employees for information and suggestions. This practice makes sure that everybody within an organization is involved. We earlier highlighted how important it is that organizations involve everybody in key decision making. A management team will be able to execute policies much more effectively when employees are involved with the formulation of policies from the start. They will be more familiar with the challenges and issues one can face while executing decisions. This will foster a work environment where there is respect for everybody's opinion. In addition to mutual respect, openness and a willingness to change are required to establish effective communication. Moreover, management can achieve effective communication by providing accurate information, clarifying the responsibilities of each team, and establishing an effective system for lodging and responding to complaints. Organizations can achieve this through the use of modern information technology. Based on the discussions presented here, we would like to emphasize that effective communication is essential for successful QMS implementation.

Apart from the key CSFs that have been discussed in this chapter, there are many other CSFs, such as project prioritization and selection, effective project management, organizational infrastructure, accountability of sponsors and champions, and the selection of an implementation partner, that an organization can look at during QMS implementation. The understanding of these CSFs is important not only from a QMS implementation viewpoint, but also from the point of view of organizational strategy.

7.4 Awareness of Some Barriers to QMS Implementation

We are now well aware of some of the major critical success factors for successful QMS implementation. Apart from an awareness of the critical success factors, however, it is also important to know what the barriers to QMS are. In fact, if a closer consideration is given, we find that most of the barriers are linked to the limitations/shortcomings of CSFs. We have emphasized that a committed and visionary leadership is key to a successful QMS implementation. However, a lack of adept leadership quality can make top management too much reliant on the middle management for guidance. We are well informed at this stage that CI efforts involve the participation of all employees of an organization in decision making, and CI normally means changing the culture that middle management has created. This can thus be one of the key barriers to implementation, as a lack of clear vision from senior leaders will lead to resistance from the middle management, since they see QMSs as a challenge to their authority. Another barrier to QMS implementation is a lack of trust between the management and employees, and due to this mistrust, employees often fail to put in enough of the required efforts. The lack of appropriate rewards and self-motivation, poor communication, and a paucity of teamwork also act as barriers to QMS implementation. We have earlier discussed that organizations need to adapt a culture that supports CI, and so a lack of a CI-supported culture is another barrier to QMS implementation. Moreover, a failure to change the organizational philosophy as per the need of the time can also be a significant barrier to advancement.

Research (Fletcher, 1999) also indicates that leaders must avoid quick-fix strategies and self-absorption for survival and must take a long-term view of empowerment that involves the commitment of the top management team to world-class excellence and the full utilization of the employees through work teams and enablement. The absence of continuous training and education of employees also hinders the QMS implementation. A lack of a process-driven focus can be a significant barrier for organizations. Organizations need to monitor continuously, effectively measure, and improve their processes. The resources requirement was well highlighted in Chapter 6, and insufficient resources act as another barrier. Apart from these indiscriminate hiring practices, ego battles among employees and management, inadequate knowledge or understanding of QMS, improper planning, a poor process improvement agenda, a short-term focus, an inability to build a learning organization, inadequate attention to customers, employee resistance, and no attempt to identify the barriers to change are some of the other challenges to a successful QMS implementation that need management's focus. Therefore, any organization willing to design, implement, or improve a QMS successfully must be aware of the different factors that can act as barriers and must continuously put some effort into eradicating them in order to make the organization a successful one. In the next section we focus our discussion on the significance of managing change in an organization.

7.5 Managing Change

There is a famous saying by Charles Kettering: "The world hates change, yet it is the only thing that has brought progress." This saying holds very true in the modern business world. Most organizations find it hard to change, whether this change is about a shift in the working practices, the organizational culture, or in management practices. We have also argued earlier that organizations need to change or adapt to change if they want to survive and be successful in this competitive world. Particularly, while discussing the introduction of a quality management system (QMS) in organizations, we have stated that one of the critical success factors is to build a quality improvement-supported organizational culture. But it is a well-known fact that instituting change in an organization is not as easily done as it is said, and executive management will face strong resistance to change from their employees. While implementing QMS, management usually are left with no other alternative than to make certain changes internally and simultaneously counter the challenges posed by the external environment. Thus, managing change is equally important as developing or designing a QMS.

Management will face resistance to change while implementing a QMS, as any change in processes, culture, management style, or working practices is not easily accepted. Research (Duschinsky, 2009) indicates that in order to survive and progress, organizations must be able to change in response to external issues of survival (i.e., increasing competitiveness) and internal issues of integration (i.e., QMS implementation). It was discussed earlier that organizations are a blend of several different cultures, and to effectively bring about a change, it is necessary to understand the impact that this combination of cultures has on the organization. The resulting diversity of perspectives due to this cultural combination can either help or hinder the organization in its change efforts. We need to emphasize here that if culture is nurtured properly, it can alone lead to a successful change initiative. On the other hand, cultural misunderstanding can even undermine a simple attempt at change. The organizational change is normally required when the existing settings do not favor organizations. The change may involve finding a solution to an existing problem resulting from the underutilization of employees or ensuring that the desired change initiatives are well integrated into the organizational structure.

To successfully handle the change, management must first identify the change agents and the organization's strategic objectives. This must be successfully communicated to all the employees and responsibilities must be assigned. Based on the data collected from various sources (i.e., interviews, surveys, observations, and assessments), the driving and restraining forces must be identified and a readiness assessment of change initiatives must be carried out. The next step is to identify strategic objectives for change and strategies for integrating diversity into change initiatives. This refers to managing processes from beginning to end, clarifying change objectives, and planning any training needs. Thereafter, management should put efforts toward minimizing any conflict that may arise due to organizational change and embed a new culture by training their employees in team building, diversity awareness, process reengineering, and mapping. Subsequently, an organization should measure and assess the accomplishment of goals and objectives. Finally, management must monitor the current strategies by establishing a channel of feedback and plan a future strategy for QMS implementation. If the management follows these suggestions while implementing QMS, the challenges associated with managing change will not be significant.

7.6 Linking to Selection Methodology

This chapter has explained some of the key CSFs for QMS implementation. Organizations need to understand how these CSFs can assist them, and if they fail to recognize their value, how this can jeopardize their QMS implementation processes. Now we would like to relate the CSFs discussed in this chapter with the selection methodology that we proposed in Chapter 6. The proposed selection methodology suggested four key activities as part of the selection of the right models, methods, and tools required to effectively design, implement, or improve an organization's QMS; these are

1. An evaluation of the need to implement the previously shortlisted models, methods, or tools
2. A cost–benefits analysis
3. An evaluation to find out whether it has the resources needed to implement the shortlisted models, methods, or tools
4. An evaluation to find out whether it has the capabilities needed to implement the shortlisted models, methods, or tools

If we refer to the CSFs that we discussed earlier in this chapter, we can visualize a close link between these criteria and the various CSFs required for the successful QMS implementation. For example, to evaluate and select the right models, methods, or tools and to perform a cost–benefit analysis, an organization must have a strong and committed leadership who can make effective decisions. Unless leaders are good visionaries, possess honed analytical skills, and are able to sense the changes happening in the internal and external environment, it would be quite hard to choose the right approaches that can complement the organization's strength. Leaders alone cannot resolve all of the issues unless they have a motivated and participative labor force that is able to work as a team toward the same organizational aim and objectives. We have already discussed how an organization can build such a participative team through training, empowering, and instituting a culture of CI. The third selection criterion focuses on the understanding of the resources that are required to implement the shortlisted approaches. An organization's resources lie in their intangible and tangible assets, such as production facilities, raw materials, cultures, technological knowledge, patents, and human capital. While discussing the CSFs we have clearly identified the importance of an organizational culture that supports CI and emphasized that the management culture should be guided by fact and not by experience or feelings. We also highlighted the need to visualize, understand, and improve processes and suggested that an organization should be process oriented. Finally, the last criterion was about developing the organizational capabilities needed to implement the shortlisted approaches. The CSFs discussed in this chapter address this criterion, as we suggested that organizations must focus on developing IT competence, empowering and training employees, building a participative workforce, establishing effective communication, and building a continuous improvement-focused culture. Thus, we can see that our proposed selection criteria link very well with the CSFs. Organizations must be able to implement QMS successfully if they recognize the importance of the CSFs as outlined and discussed in this chapter.

7.7 Summary

In this chapter we have focused on the critical success factors (CSFs) that are vital for QMS implementation. We started by highlighting the fact that QMS implementation is not at all straightforward, and management often has to struggle hard due to the substantial challenges posed by several factors during its implementation. We then identified the role of CSFs in the design, implementation, or improvement of a QMS. In particular, we discussed the five important CSFs: a committed leadership, a motivated labor force, a process-oriented focus, an organizational culture-supporting CI, and effective communication. All these CSFs are discussed in detail together with an emphasis on how organizations can develop them. We have also identified some barriers to QMS implementation that are very closely linked with these CSFs. Finally, we have put an emphasis on managing change before linking the selection criteria presented in Chapter 6 with the CSFs highlighted in this chapter. Therefore, this chapter provides managers and practitioners with an understanding of the critical success factors essential for QMS implementation, and in doing so, it also highlights the importance of managing change that all together contributes to a successful QMS.

7.7.1 Key Points to Remember

- Management has to overcome several challenges in order to implement QMS successfully in an organization.
- Organizations trying to implement a quality improvement framework must continuously seek to identify critical success factors (CSFs).
- A strong, committed leadership and good decision-making skills are vital for a successful implementation of a QMS.
- Organizations need to have both a strong and committed leadership as well as a highly motivated and committed workforce. A lack of motivation and commitment among employees and top management can act as a hindrance to QMS implementation.
- An understanding of processes is crucial since managing quality within the organizations is very much dependent on the way the organizations manage their processes. Thus, an organization must establish a process-oriented culture.
- To ensure that organizations continue to follow the right path without any obstacles, an organizational culture that supports continuous improvement is essential.
- There is a strong relationship between good communication and successful quality implementation. Thus, an organization must develop an effective system of communication.
- There are several barriers to QMS implementation, and organizations must overcome these barriers.
- Managing change is as important as instituting a strong supportive culture.

References

Coronado, R. B., and Antony, J. (2002). Critical success factors for the successful implementation of Six Sigma projects in organisations. *TQM Magazine*, Vol. 14, No. 2, pp. 92–99.

Duschinsky, P. (2009). *The change equation: How to identify, quantify and overcome the real barriers to organisational change*. Management Books 2000, Cirencester, UK.

Fletcher, M. (1999). The effects of internal communication, leadership, and team performance on successful service quality implementation: A South African perspective. *Team Performance Management*, Vol. 5, No. 5, pp. 150–163(14).

Ocholi, S. A. (1998). The role of effective communication in the application of total quality management (TQM). *Nigerian Journal of Management Research*, Vol. 1, No. 1, pp. 197–207.

O'Reilly III, C. A., and Chatman, J. A. (1996). Culture as social control: Corporations, cults, and commitment. In Staw, B. M., and Cummings, L. L. (Eds.), *Research in organizational behavior*, Vol. 18, pp. 157–200. JAI Press, Greenwich, CT.

Thite, M. (2000). Leadership styles in information technology projects. *International Journal of Project Management*, Vol. 18, pp. 235–241.

Wang, E., Chou, H. W., and Jiang, J. (2005). The impacts of charismatic leadership style on team cohesiveness and overall performance during ERP implementation. *International Journal of Project Management*, Vol. 23, No. 3, pp. 173–180.

Further Suggested Reading

Amar, K., and Zain, Z. M. (2002). Barriers to implementing TQM in Indonesian manufacturing organizations. *TQM Magazine*, Vol. 14, No. 6, pp. 367–372.

Deter, J. R., Schroeder, R. G., and Mauriel, J. J. (2000). A framework for linking culture and improvement initiatives in organizations. *Academy of Management Review*, Vol. 25, No. 4, pp. 850–863.

Developing a culture of continuous improvement: An overview of organisational culture: how it forms and how you may work to change it over time, an Amnis Executive Insight Paper. (2012). Amnis, Surrey, UK.

Henderson, K., and Evans, J. (2000). Successful implementation of Six Sigma: Benchmarking General Electric company. *Benchmarking and International Journal*, Vol. 7, No. 4, pp. 260–281.

Pascale, R. T., and Sternin, J. (2005). Your company's secret change agents. *Harvard Business Review*, May 2005, pp. 1–10.

Slocum, J. W. (2000). Leadership and decision making process. *Organizational Dynamics*, Vol. 28, No. 4, pp. 82–94.

Wickisier, E. L. (1997). The paradox of empowerment—a case study. *Empowerment in Organizations*, Vol. 5, No. 4, pp. 213–219.

Chapter 8

QMS and Business Processes Evaluation

8.1 Introduction

The design or improvement of a quality management system (QMS) and business processes does not stop with the selection and implementation of the right business and quality improvement models, methods, and tools. Once integrated into the QMS or improvement plan, they have to be monitored and evaluated to determine their relevance and the benefits they provide to the organization. Measuring the progress of QMS and business processes is a means of conducting follow-up evaluations to determine whether they are still benefiting the organization. In this chapter we propose and adapt the diagnostic methodology presented in Chapter 4 as an approach to also determine whether improvement activities have benefited the organization's QMS and business processes and whether these benefits are being sustained over the long run. The chapter begins by discussing the importance of follow-up activities in continuous improvement (CI) and the specific information that they can provide to organizations to support the continuous success of improvement initiatives. Then, we introduce a follow-up evaluation method that consists of replicating the diagnostic methodology through specifically adapted versions of the maturity diagnostic instrument (MDI), a self-assessment process and quality management audit. Finally, we conclude the chapter with a brief discussion aimed at making organizations aware of the importance of effectively managing their CI experience and knowledge as a strategy for achieving sustainable business excellence.

8.2 Follow-Up Activities

The key to a successful QMS and business processes is the completion of dedicated follow-up activities carried out by the organization. The follow-up of improvement activities demonstrates that the organization supports and is serious about CI, and that top management can serve as a resource for staff members if they require assistance with their CI tasks and activities. It is therefore important for the organization to develop a culture through which to not only carry out business and quality improvement activities but also follow them up to ensure that they are sustained and provide the expected results. In particular, follow-up activities will help an organization to answer the following questions:

- How effective have the selection and implementation of the selected business and quality improvements models, methods, and tools been?
- Have these models, methods, and tools been effectively sustained over time?
- Has the implementation of these approaches benefited the organization's QMS or business processes?
- Has this benefit been sustained over the long run?
- Do the same issues highlighted by the QMS and business process diagnostic still exist?
- Have new issues that need to be addressed emerged?

The follow-up process should mainly consist of measuring the effectiveness and progress of the QMS and business processes after they have been designed or subjected to any improvement initiatives. By answering the above questions, an organization will be able to validate or modify its improvement plan accordingly. This is because the follow-up process will provide the organization with an early warning system to detect unwanted deviations in the effectiveness of its QMS or business processes so that immediate and appropriate corrective actions can be taken.

Ideally, top management should assign the follow-up measuring task to the same team that carried out the diagnosis of the organization's QMS and business processes. This will allow the follow-up activities to be carried out more effectively and efficiently, as the evaluating team would already be familiar with the organization's activities and processes as well as the evaluation method we propose and present in the following section. This team would therefore require minimum, or no, training to carry out the evaluation. Alternatively, a new follow-up team can be formed, although in this case team members will have to receive the appropriate training and will also require some time to get familiar with the organization.

Top management should also establish a follow-up routine that consists of regular evaluations of the QMS and business processes. The frequency of evaluations should be determined based on the maturity of the QMS and the results of the self-assessment process and quality management audit. Clearly, organizations with more mature QMSs and effective business processes that comply with quality standards would require less frequent follow-up evaluations than those that present poor performances. If the QMS or business processes are not functioning as expected, follow-up evaluations will consume organizational resources, particularly staff time to carry out the assessment, analyze the results, and propose and implement the corresponding corrective actions. For this reason, the availability of resources to perform the follow-up process will also play an important role in determining the frequency with which this activity is carried out. Undoubtedly, follow-up evaluations will benefit the organization, but at the same time they will represent a cost. It is for this reason that organizations will also need to define an economically healthy and cost-effective number and frequency of follow-up activities and evaluations to ensure that the cost of these does not exceed their expected benefit.

8.3 Follow-Up Evaluation Method

We recommend employing the same diagnostic methodology proposed in Chapter 4 as a method for measuring the effectiveness and progress of the QMS and business processes after they have been designed or subjected to any improvement initiatives. As previously discussed, the diagnostic methodology can be used to present not only a picture of the original state of the organization's QMS and business processes, but also an after improvements picture. Updating the data in the diagnostic methodology after the design or improvements have taken place allows the results to be compared against those of the organization's original state, as illustrated in Figure 8.1. This will help the organization to answer the questions presented in the last section and thus determine whether any progress has been made.

Similar to the diagnostic methodology, the evaluation method we propose consists of performing the same three assessments, which include:

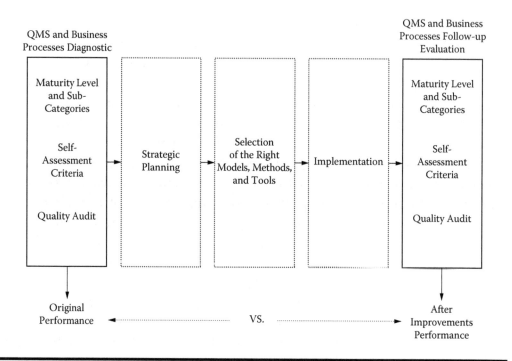

Figure 8.1 Diagnostic methodology vs. follow-up evaluation—comparison of results.

- A maturity evaluation using an adapted version of the MDI
- An evaluation of the organization's business processes by performing a self-assessment process using a business excellence model (BEM)
- A first-party audit

8.3.1 Defining the QMS Maturity for Follow-Up Evaluations

Progress in the maturity of a QMS and the subcategories evaluated by the MDI can provide a clear indication as to whether the effectiveness of a QMS has improved after the implementation of the selected business and quality improvement approaches. Table 8.1 presents an adapted version of the MDI that can be used by organizations to carry out and record follow-up maturity evaluations. Similar to the original MDI, a score of 1 to 7 has to be assigned based on the evaluation team's perception regarding the position of the company in relation to every one of the subcategories after improvements. In the "original performance" (OP) column the initial score assigned to every subcategory during the initial maturity diagnostic should be recorded. Then, after each follow-up evaluation (e.g., E1, E2, E3), the scores assigned to each subcategory should also be recorded and compared with the OP score and

Table 8.1   Adapted Version of the MDI for Follow-Up Evaluations

Subcategory	Original Performance	After Improvements Performance			
	From 1 (strongly agree) to 7 (strongly disagree)				
	OP	E1	E2	E3	En
1. Quality improvement (QI) initiatives *are not* only carried out to achieve ISO 9000 registration or comply with customer requirements.					
2. Initial enthusiasm after implementing a quality management system (QMS) or QI program *does not* fade over time.					
3. Organization holds an ISO 9000 certification (or is close to obtaining it).					
4. Organization recognizes that the effective implementation of a QMS requires cultural change.					
5. Organization has a culture where quality *is not* dependent on the commitment and drive of a limited number of individuals.					
6. A total integration of continuous improvement (CI) and business strategy to delight customers exists.					
7. Organization *does not* only apply quality management (QM) tools and techniques due to customers' presence, monitoring, and pressure.					
8. Organization *has not* expressed disappointment about the current QMS.					
9. Organization employs a selection of quality management tools (e.g., statistical process control (SPC), quality circle (QC), failure mode and effects analysis (FMEA), mistake proofing, quality improvement groups).					
10. Organization recognizes the importance of customer-focused CI.					

Table 8.1 *(Continued)* Adapted Version of the MDI for Follow-Up Evaluations

	Original Performance	After Improvements Performance			
Subcategory	From 1 (strongly agree) to 7 (strongly disagree)				
	OP	E1	E2	E3	En
11. All employees are involved in CI.					
12. Organization's purpose and values are defined and communicated at all levels.					
13. Not only does the quality department drive the QMS and maintain ISO certification, but all staff participate and have concern for quality.					
14. Organization *is not* susceptible to the adoption of the latest QM fads.					
15. Organization *does not* tend to look for the latest QI approaches/tools for a "quick fix."					
16. Senior management shows commitment toward QI through both leadership and personal actions.					
17. A number of successful organizational changes have been made.					
18. Organization has developed and applied a unique success model.					
19. Success of quality initiatives *is not* linked to the success of external audits only.					
20. Management teams *do not* try a variety of approaches in response to the latest QM fads.					
21. All senior management members are committed to the organization's QMS.					
22. Organization has formulated a quality strategy and implemented, at least, a good portion of it.					
23. Business procedures and processes are efficient and responsive to customer needs.					

Table 8.1 *(Continued)*   **Adapted Version of the MDI for Follow-Up Evaluations**

Subcategory	Original Performance	After Improvements Performance			
	From 1 (strongly agree) to 7 (strongly disagree)				
	OP	E1	E2	E3	En
24. Organization places a positive value on internal and external relationships (e.g., with employees, customers).					
25. QM *is not* considered a contractual requirement and an added cost.					
26. Senior management *does not* assume that CI occurs naturally or is self-sustained.					
27. CI efforts are concentrated not only in manufacturing/operations departments but also in other departments of the organization.					
28. A problem-solving infrastructure and a proactive QMS are in place.					
29. Process improvement results are measurable and carried out through effective cross-functional management.					
30. Organization works in partnership with stakeholders.					
31. Priority is given to QI in terms of time and allocation of resources.					
32. Organization has adopted different quality philosophies (e.g., Deming, Crosby, Juran, SPC, International Organization for Standardization (ISO), Total Quality Management (TQM), Six Sigma).					
33. A QMS exists and the data it provides are used to their full potential.					
34. A long-term and company-wide education/training program is in place.					
35. Strategic benchmarking is practiced at all levels.					

Table 8.1 *(Continued)*   Adapted Version of the MDI for Follow-Up Evaluations

	Original Performance	After Improvements Performance			
		From 1 (strongly agree) to 7 (strongly disagree)			
Subcategory	OP	E1	E2	E3	En
36. QMS helps to identify opportunities to improve the ability of the company to satisfy its customers.					
37. Corrective actions *are not* only taken in response to customer complaints.					
38. Continuous improvement is perceived as a strategy, not as a program only.					
39. Long-term results in all organizational aspects (as opposed to short-term results regarding product output and quality only) are expected.					
40. Individual staff carry out improvement activities within their own spheres of influence and on their own initiative.					
41. A system for internal and external performance measurement is in place.					
42. Organization is constantly looking to identify new/more products, services, or characteristics that will increase customer satisfaction.					
43. Support to solve problems *is not* based on their impact on sales/turnover only.					
44. A plan for effectively deploying a QMS exists.					
45. Processes *do not* have considerable potential for improvement.					
46. Importance of staff involvement in CI is recognized, communicated, and celebrated.					
47. Employees at all levels reflect a participate culture.					
48. A QI culture is no longer dependent on top-down drives, but it is also driven laterally through the whole organization.					

Table 8.1 *(Continued)*   **Adapted Version of the MDI for Follow-Up Evaluations**

Subcategory	Original Performance	After Improvements Performance			
	From 1 (strongly agree) to 7 (strongly disagree)				
	OP	E1	E2	E3	En
49. Quality of design has a high priority.					
50. Management *is not* oversusceptible to outside intervention and *does not* easily get distracted by the latest QM and CI fads.					
51. All parts of the organization believe that the current QMS is effective.					
52. Benchmarking studies have been initiated and the results used for CI.					
53. Management practices a culture of empowerment.					
54. The vision of the entire organization is aligned to the voice of the customer.					
55. Organization has made an acceptable investment on quality education and training.					
56. Quality department has a high status within the organization.					
57. Momentum of improvement initiatives is easy to sustain.					
58. Organization has QI champions among some senior management members.					
59. Current QMS is sincerely viewed by all employees as a way of managing the business to satisfy and delight customers, both internal and external.					
60. Total quality is the organization's "way of life" and "way of doing business."					
61. Senior management takes responsibility for CI/QI activities.					
62. The "born and died" of improvement teams *is not* a constant phenomenon.					

Table 8.1 *(Continued)*   Adapted Version of the MDI for Follow-Up Evaluations

	Original Performance	After Improvements Performance			
		From 1 (strongly agree) to 7 (strongly disagree)			
Subcategory	*OP*	*E1*	*E2*	*E3*	*En*
63. Training on quality tools is aimed at persons who can influence their further application.					
64. Trust between all levels of the organization exists.					
65. Perception of stakeholders of the company's performance is surveyed and acted on to drive improvement actions.					
66. Quality values are fully understood and shared by employees, customers, and suppliers.					
67. Organization has had positive previous experience with ISO, TQM, or other quality management approaches.					
68. Cultural changes have taken place after the implementation of CI/QI programs.					
69. Quality tools and techniques are implemented strategically and not only reactively and when necessary.					
70. There is low preoccupation with numbers (e.g., financial measures).					
71. Results of improvement projects are effectively utilized.					
72. Each person in the organization is committed, in an almost natural way, to seek opportunities for improvement.					
73. There *is not* an overwhelming emphasis on the achievement of financial measures.					
74. Appropriate knowledge of the current QMS exists.					

Table 8.1 *(Continued)* **Adapted Version of the MDI for Follow-Up Evaluations**

	Original Performance	After Improvements Performance			
		From 1 (strongly agree) to 7 (strongly disagree)			
Subcategory	OP	E1	E2	E3	En
75. Meeting output targets *is not* the only key priority for the majority of managers; there are no conflicts between the production/ operations department and the quality department.					
76. QI drives and direction *do not* rely only on a small number of individuals.					
77. All things are done right the first time.					
78. Dependability is emphasized throughout the organization.					
79. There is a long-term plan for corrective actions for reoccurrence of problems.					
80. Self-assessment is performed and improvements identified are addressed.					
81. The organization has a flexible QMS not only designed to fulfill customer regulations.					
82. If key directors/managers/individuals leave, business mergers occur, organizational restructuring takes place, etc., there *is no* danger of losing momentum or failure in terms of QM/QI initiatives.					
83. QMS is effective and it does help to identify opportunities to improve the ability of the company to satisfy its customers.					
84. Waste is not tolerated.					

the scores of previous evaluations. This will show whether an organization has attained and sustained any improvements. For example, if an OP score of 4 is obtained in a specific subcategory during the initial maturity diagnostic, and then scores of 3 in E1 and 2 in E2 are assigned, this would indicate a steady improvement in that subcategory. On the other hand, if OP = 4,

Table 8.2   Maturity Recoding Form for Follow-Up Evaluations

Original Maturity Level	After Improvements Maturity Level			
OML	E1	E2	E3	En

E1 = 2, and E2 = 3, this would indicate that an improvement has been achieved but not sustained.

Every time a follow-up evaluation is carried out and after the scores have been recorded for each one of the subcategories, the same procedure followed for the original MDI must be performed. This refers to the procedure for transferring the scores to their corresponding columns in the scoring table (Table 4.2). Next, they need to be added, and the result of each sum divided by 14 to obtain comparable scores. Like in the original MDI, the highest score will indicate the overall status of quality maturity and category (e.g., "uncommitted," "drifters," etc.) of the organization. Finally, the maturity level should be recorded in Table 8.2 for comparative and historical purposes. Naturally, a move from one category to a more mature one will indicate an improvement in the effectiveness of the organization's QMS.

8.3.2 Follow-Up Evaluations for Business Processes

In addition to evaluating the maturity progress of its QMS, it is also important for an organization to focus on assessing whether its business processes have progressed after the deployment of any improvement initiative. Diligence in following up on this progress will provide an organization with information about whether the strengths identified through the self-assessment process have been maintained and the weaknesses improved on. To do this, we suggest performing a follow-up self-assessment, similar to the one carried out as part of the diagnostic, and following our best-practice approach for conducting a self-assessment process presented in Section 4.3.1.

The follow-up self-assessment should be carried out using the same BEM and evaluating the same criteria and subcriteria as in the initial self-assessment. Table 8.3 provides a form for organizations to use to carry out and record the results of the follow-up self-assessments. Table 8.3 has been specifically designed for an organization using the EFQM model; organizations using a different model will need to adapt the specific criteria and subcriteria of such a model to this format. Here, the key is to list all the same criteria and subcriteria previously used for evaluation in the initial

Table 8.3 Form for the Follow-Up of Self-Assessments

	Available Points	OP	E1	E2	E3	En
ENABLERS						
Leadership	**100**					
Visible involvement in leading TQ	16.66					
A consistent TQ culture	16.66					
Timely recognition and appreciation of the efforts and successes of individuals and teams	16.66					
Support of TQ by provision of appropriate resources and assistance	16.66					
Involvement with customers and suppliers	16.66					
Active promotion of TQ outside the organization	16.66					
Policy and strategy	**80**					
How policy and strategy are formulated on the concept of TQ	16					
How policy and strategy are based on information that is relevant and comprehensive	16					
How policy and strategy are implemented throughout the organization	16					
How policy and strategy are communicated internally and externally	16					
How policy and strategy are regularly updated and improved	16					
People management	**90**					
How people resources are planned and improved	18					
How the skills and capabilities of the people are preserved and developed through recruitment, training, and career progression	18					

Table 8.3 *(Continued)* **Form for the Follow-Up of Self-Assessments**

	Available Points	OP	E1	E2	E3	En
How people and teams agree on targets and continuously review performance	18					
How the involvement of everyone in CI is promoted and people are empowered to take appropriate action	18					
How effective top-down, bottom-up, and lateral communication is achieved	18					
Resources	**90**					
Financial resources	22.5					
Information resources	22.5					
Suppliers, material, buildings, and equipment	22.5					
The application of technology	22.5					
Processes	**140**					
How processes critical to the success of the business are indentified	28					
How the organization systematically manages its processes	28					
How processes are reviewed and targets are set for improvement	28					
How the organization stimulates innovation and creativity in process improvement	28					
How the organization implements process changes and evaluates the benefits	28					
RESULTS						
Customer satisfaction	**200**					
The customers' perception of the organization's products, services, and customer relationships	150					
Additional measures relating to satisfaction of the organization's customers	50					

Table 8.3 *(Continued)*   Form for the Follow-Up of Self-Assessments

	Available Points	OP	E1	E2	E3	En
People satisfaction	**90**					
The peoples' perception of the organization	67.5					
Additional measures relating to people satisfaction	22.5					
Impact on society	**60**					
The perception of the community at large of the organization's success in satisfying the needs and expectations of the community at large	15					
Additional measures relating to the organization's impact on society	45					
Business results	**150**					
Financial measures of the organization's success	75					
Nonfinancial measures of the organization's success	75					

self-assessment, and then create some extra columns where the performance of the business processes can be recorded for each follow-up evaluation (e.g., E1, E2, E3). Table 8.3 lists all of the 9 criteria and 33 subcriteria that comprise the EFQM model, as well as the specific number of points available for each one of them. In the "OP" column, the scores assigned to every criterion and subcriterion during the initial self-assessment should be recorded. The rest of the columns should be employed to record the scores assigned to every criterion and subcriterion while the different follow-up self-assessments are performed. An increase in a particular score, for example, from OP to E1 or from E1 to E2, would obviously indicate an improvement in that category or subcategory. On the other hand, a decrease in the score will indicate that the improvement changes carried out have not benefited the progress of the business process, but made it worse. Similarly, as with the follow-up maturity evaluations, a follow-up assessment can also indicate whether the improvements achieved in the organization's business processes have been sustained.

8.3.3 Follow-Up Evaluations of Quality Management Audits

Having a mature QMS and effective business processes does not necessarily mean that an organization's products, services, or processes will fully comply with the requirements of its customers, suppliers, partners, collaborators, industry sector, or government regulations. Thus, quality management audits play a key role in ensuring the effectiveness of a QMS and in identifying any procedures that may not conform to specifications. Once those noncompliance procedures have been subjected to improvement initiatives, it is vital for an organization to find out whether these initiatives have provided the expected results. If no noncompliance quality assurance procedures were highlighted, then it is still important for an organization to know that these have not deviated, and thus still comply with the corresponding regulations. It is for these reasons that in addition to the maturity and self-assessment follow-ups, we also suggest performing follow-up quality management audits to validate progress actions and the effective implementation of business and quality improvement approaches. In Section 4.4 we presented a procedure for conducting quality management audits during the QMS and business process diagnostic stage. This same procedure can also be followed to perform follow-up audits.

The audit forms used to assess the compliance of organizational quality procedures vary greatly in industry. However, in Table 8.4 we provide a generic form that we have adapted for the purpose of comparing the results of the initial quality audit with those of subsequent follow-up audit evaluations.

Similar to the quality maturity and self-assessment follow-up forms, in the "original performance" (OP) column the evaluation code (see at the bottom of Table 8.4) for every quality procedure audited during the diagnostic stage must be recorded. Columns for follow-up audits (e.g., E1, E2, E3) should be filled with the evaluation codes assigned to each procedure. In this way, different performances can be easily compared to find out whether any progress has been achieved in improving nonconformances or if the quality assurance procedures still satisfy the corresponding requirements.

8.4 Lessons Learned and the Management of Knowledge for Business Excellence

In this book we have focused on providing a series of methodologies and recommendations for effectively designing or improving an organization's QMS and core business processes. The adequate functioning of these two

Table 8.4   Form for the Follow-Up of Quality Management Audits

Insert company's logo here	Quality audit checklist for _____		Issue:		Revision:	Page __ of __
	Reviewed by:		Approved by:			
Doc. code:	Report no.:		Auditor(s):		Date:	

Internal (First-Party) Audit

Code of Procedure Checked	Audit Question	Code					Observations
		OP	E1	E2	E3	En	

Evaluation code:	AC = Acceptable	IR = Improvement required	UN = Unacceptable	N/A = Not applicable

key organizational elements is essential to a company in maintaining a competitive edge over its rivals and meeting the expectations of its customers. Once appropriate actions have been taken to design or improve an organization's QMS and core business processes, and positive results have been achieved, the fundamental challenge then becomes how to sustain and constantly repeat such success. This is where an organization has to make sure that key experiences acquired during the whole improvement process are shared through the relevant departments and members of the organization. This will ensure that good practices are repeated and institutionalized and that the same problems do not occur again. Unfortunately, this is not always easy to do, nor is it a common practice in industry. For instance, some statistics indicate that 80% of all quality problems in the manufacturing industry are recurring issues. In other words, these are errors that have occurred before and were fixed, yet the lessons learned from such errors and their solutions were not remembered or communicated to other groups so that preventive actions could be taken. The explanation on the part of managers for this phenomenon included the "inability to manage lessons learned and best practices" and "poor communication between engineering and manufacturing." It is therefore important for organizations to transform their CI experiences into lessons learned and make them part of their improvement plan and QMS so that they are readily available to the departments and staff involved in CI projects.

Knowledge management and CI are complementary practices that, when combined, can create a synergy to assist organizations in their journey toward excellence. In this section we have tried to highlight this fact to make organizations aware of the need for creating and implementing adequate mechanisms for the effective management and communication of their improvement experiences and knowledge. The area of knowledge management has received a lot of attention over the last two decades, as it has been recognized by academic researchers as one of the pillars for business excellence. This has contributed to the development of various models, or even computer software, that organizations can adopt to systematically identify, document, and benefit from lessons learned. In the further suggested reading section at the end of this chapter we included some reference texts that can be consulted to guide an organization in its quest for an effective management of its CI knowledge. Alternatively, we also recommend that organizations seek professional advice and guidance from professional institutions, local universities, or consultants regarding the implementation of knowledge management practices.

8.5 Summary

In this chapter we have discussed and emphasized the importance of follow-up activities after the QMS or business processes have been designed or subjected to any improvement activities. Specifically, we argue that follow-up evaluations play an essential role in an organization in terms of discovering whether the improvement actions taken have delivered the expected results and been maintained over the long term. To carry out the follow-up evaluations, we proposed the use of the QMS and business processes diagnostic methodology presented in Chapter 4. In this way, we have adapted the diagnostic methodology to be replicated as a follow-up method. This will aid organizations in carrying out more effective and efficient follow-ups, since they will already have experience and practice applying the maturity evaluation, self-assessment process, and quality management audits contained in the diagnostic methodology.

The follow-up evaluation method we proposed consists of replicating the maturity evaluation, self-assessment process, and quality management audit following the same guidelines we provided in Chapter 4. The difference lies in the recording of the follow-up results, for which we have adapted and provided some specific forms and guidelines. These forms will allow an easy comparison to be made between the original performance of the QMS and business processes obtained during the diagnostic stage and their performance during subsequent follow-up evaluations. The comparisons will provide a clear picture as to whether

- The selection and implementation of the chosen business and quality improvements models, methods and tools have been effective and sustained over time
- The implementation of these approaches has benefited the organization's QMS or business processes, and whether this benefit has been sustained over the long run
- The same issues highlighted during the QMS and business processes diagnostic still exist
- New issues that need to be addressed have emerged

Finally, in this chapter we have also briefly discussed the importance of learning from CI experiences and making that knowledge readily available to the departments and individuals involved in continuous improvement projects. Learning is the acquisition of new knowledge about the way the world works; this must occur in CI if it is to provide the intended benefits. The effective share and management of this learning is known as knowledge management (KM). KM is currently considered one of the pillars of business excellence; for this reason, we have also highlighted in this chapter the need for organizations to adopt this practice as part of the effective management of the organization's business. In Chapter 9, we will discuss the behaviors, attitudes, actions, and activities required of an organization in order to institutionalize a culture committed to quality and the effective management of its core business processes.

8.5.1 Key Points to Remember

■ Once the business and quality improvement models, methods, and tools have been selected and integrated into the QMS or improvement plan, follow-up activities and evaluations have to be established to determine whether they are benefiting the organization.
■ Follow-up evaluations can be carried out using the method we proposed in this chapter, namely, the adaptation of the QMS and business processes diagnostic methodology previously presented in Chapter 4.
■ The difference between the methodology presented in Chapter 4 and its adaptation as a follow-up method lies in the recording of the follow-up results, for which we have provided some specific forms and guidelines.
■ To ensure that success is sustained and constantly repeated, an organization has to ensure that key experiences acquired during an improvement initiative are shared through the relevant departments and members of the organization.

Further Suggested Reading

Armistead, C. (1999). Knowledge management and process performance. *Journal of Knowledge Management*, Vol. 3, No. 2, pp. 143–154.
Basu, R., and Wright, J. N. (2003). *Quality beyond Six Sigma*. Elsevier, Butterworth Heinemann, Oxford.
Cobb, C. G. (2003). *From quality to business excellence—a system approach to management*. ASQ Quality Press, Milwaukee.
Evan, J. R., and Lindsay, W. M. (2002). *The management and control of quality*. South-Western Thomson Learning, Cincinnati, OH.
Harrington, H. J., and Voehl, F. (2007). *Knowledge management excellence*. Paton Press, Chico, CA.
Van Geenhuizen, M., Trzmielak, D., Gibson, D. V., and Urbaniak, M. (2009). *Value added partnering and innovation in a changing world (international series on technology policy and innovation)*. Purdue University Press, West Lafayette, IN.

Chapter 9

Beyond Quality Management Systems

9.1 Brief Summary

Since we have covered some of the topics related to quality management systems (QMSs), you should have a good understanding of the QMS along with some of the main quality methods and tools. First, we reviewed the general issues related to QMSs and the importance of such systems in the competitive business environment and their alignment with business strategies. Next, we covered business excellence models (BEMs), and quality management standards as general umbrellas, to support and deploy quality methods and tools. After this, we stated the importance of having a process-oriented organization supported by a strong information technology infrastructure to automate business and run efficient and effective processes. Then, we provided a method to deploy the diagnosis of the QMS to help determine the quality maturity level of the organization. Without knowing where the organization is and where it should lead, all directions appear the same, and there is a high risk of getting lost.

That is the reason why we introduced strategic quality planning; it helps to give systematic direction to improvement programs and integrates all quality management efforts with organizational performance and business strategy. Then, decision-making skills play an important role in the selection of the right quality models, methods, and tools that should be part of the QMS. The problem here is not about how many quality models, methods, and techniques we know. The decision-making issue here is to select the most appropriate quality models, methods, and tools for your company based on the organization's needs, requirements, capabilities, and resources. From that standpoint, we have to build a strong and sustainable QMS. Until this point, everything is under control; however, when deploying the QMS or any business improvement strategy, things generally do not go as planned. So, be prepared.

That was the reason why we introduced sections on change management and awareness of the main barriers to the successful implementation of QMSs. Being a learning organization means learning from others' experiences, from past projects, and from success and failure, and understanding the causes of these successes and failures. Avoiding and overcoming problems and barriers as well as developing strong leadership are skills that managers should have and continuously improve. There is still no magic method for accomplishing this; the best organizations can do is to invest heavily in training and carefully select, develop, and retain their most talented people. Then, after the deployment of the QMS, the challenge is to measure what we have done, and to measure it precisely and systematically. Specifically, we have to compare improvement outputs with current quality strategic objectives and set any corrective actions. Continuous improvement can be accomplished by identifying areas of opportunity and prioritizing them so that we define what improvement methods to use to tackle those opportunity areas. Then we can go beyond the QMS.

9.2 Quality Management Culture: A Way of Life

You must embrace a quality management culture as a way of doing business, with open communication and the eagerness to learn from success and failure. First, your organization needs to have a good-quality maturity level to understand that the quality culture is the central environment in which business is conducted. Make sure that you set the strategic quality plan to reach the desired maturity level in a realistic time frame. Then, it is compulsory that key employees are aware and trained so that they can disseminate the vision, values, and way of working and doing business on a daily basis. They need to share the vision in the long term that characterizes excellent organizations. It is also essential to conduct an assessment to determine the organizational climate and to seek evidence that people are committed, well trained, and motivated to do their jobs and fulfill the function they currently perform. After the assessment, there could be several scenarios, and from that point, you must implement and deploy the programs that foster the desired organizational climate. Some of these programs may be related to personnel training, compensation schemes, company share participation, working environments, salaries, and recognition schemes, among others. The essential aim is to ensure that people are happy with the things they do, the salaries they get, and the environment in which they work, and that they have the resources they need to do their jobs with high-quality standards.

This is the crucial point so that people at all levels of the organization outperform what it is expected from them, and this is also the right environment in which to build a strong foundation for the quality culture. Systems, structure, and well-defined processes, along with IT infrastructure and resources, are also a must to ensure efficiency in doing business. Finally, you need to keep in mind that the QMS itself is just a means, not the destination. The challenge is to do business in a sustainable way in the long term and meet the expectations of all stakeholders, such as customers, employees, society, environment, investors, and government.

As a leader in your organization, you need to provide the elements to bring about the cultural change. Taking on this responsibility is a big challenge for any CEO or director, since it involves working and focusing strictly on the future of your organization without compromising the present. Frankly, daily management that works on the present, on the day-to-day operations, has a heavy responsibility (Figure 9.1). We are talking about people who deliver medicine, cashiers in supermarkets, receptionists answering multiple calls, and bank tellers—they deal with customers every day. They are our unknown heroes, and the performance of any organization depends

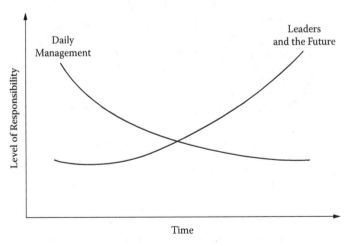

Figure 9.1   Leaders' level of responsibility.

directly on their responsibilities and duties. On the other hand, leaders work in the future. They should provide direction to the organization; their responsibility is to lead business to higher levels by making the right decisions about products, markets, infrastructure, and business models, and providing high-quality products and services. They have to overcome paradigms, things that have also been done in a particular way, and now they must change, and change quickly.

Thus, to have a company with a strong quality culture requires that you have strong leadership, lead by example, and motivate others to follow your objectives, values, and vision. Therefore, you need to provide the elements to develop a quality culture in which the entire organization is committed to the things it does, and to achieve the values of excellence and perfection.

9.3 The Never-Ending Improvement Process

The never-ending improvement process for the QMS consists of the following steps (Figure 9.2):

- Define customer needs, requirements, and expectations.
- Set specific objectives to address the customers' issues mentioned above.

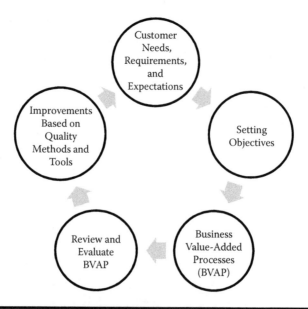

Figure 9.2 The never-ending improvement process.

- Identify value-added and business value-added processes to achieve objectives.
- Deploy processes and systematically review/evaluate them based on objectives.
- Determine quality models, methods, and tools to improve processes.
- Review customer needs, requirements, and expectations as necessary.
- Close the loop and redefine objectives if necessary.

9.3.1 Customers' Needs, Requirements, and Expectations

We began and will close this book with a reminder that customers are the *raison d'être* of any organization. It is essential to understand their needs, requirements, and expectations, and then translate them into the product and services they require. Any quality model and method needs to start from this point, followed by the design of the processes based on those requirements. Some initiatives, such as the voice of the customer (VOC) and methods that include quality function deployment (QFD) and Design for Six Sigma (DFSS), can be helpful for understanding and translating customers' needs and requirements to design.

9.3.2 Set Specific Objectives

Once the customer needs and requirements are identified, set realistic and feasible objectives for the continuous improvement plan. Those objectives have to be clear and provide the specific metrics to monitor and measure the progress.

9.3.3 Value-Added and Business Value-Added Processes

Make sure that you identify correctly the processes that add value to your business. It is very common to lose focus of the things that actually add value to a business. Then it is necessary to map those processes with a value stream mapping (VSM) technique so that you actually map the current stages of your business and the desired state of them. Any other approach, such as business process reengineering, Six Sigma, and ISO standards, is also highly valuable at this stage. Select approaches based on the organizational resources, capabilities, and needs, and make sure that you get a cost–benefit implementation.

9.3.4 Review and Evaluate Progress toward Objectives

Regularly check the progress of the business improvements. We suggest doing it monthly, quarterly, and yearly. Of course, this will depend on the size of the projects and current policies and requirements of any organization. The issue here is to not lose too much time on activities that do not add value and to concentrate resources on things that do add value.

9.3.5 Quality Methods and Tools

Make good decisions based on reviews and evaluations when selecting the quality methods and tools for process improvements. A wide range of them are provided in Chapter 6. Base your decision on your organization's needs, resources, technical feasibility, and the costs and benefits that these approaches offer. Avoid "programs du jour" and management fads.

Then, again, go to customers' needs, requirements, and expectations and close the loop. Do it systematically as many times as necessary. The journey to excellence is a never-ending process.

9.4 Becoming a World-Class Organization

This has to be one of the most important objectives for your company, and you must be committed to invest the necessary resources at all levels to reach this point. Whatever the maturity level of your QMS at this point, strategic quality planning and deployment can help you reach high-quality standards. It is a matter of the resources that your organization is committed to invest and the time frame to get to this point.

Does a QMS ensure business success? To answer this question, we should state the attributes of a successful organization:

■ Exceeds customers' expectations
■ Has strong leadership with clear objectives and a shared vision and values
■ Has effective strategic planning and overall business strategies
■ Runs processes efficiently and effectively
■ Has talented people who are motivated, well trained, and committed to stay with the organization in the long term
■ Constantly and systematically measures what it does with a robust performance measurement system

- Has robust and strong financial performance
- Integrates technologies to make business efficient and effective
- Views learning and continuous improvement not as requisites or standards but a way of life

There is strong evidence to suggest that companies that have a well-structured and developed QMS outperform their competitors. The use of a QMS is fundamental to support business performance, provide a range of benefits for business improvements, and thus positively affect the organization (Marash et al., 2004). After implementing a QMS, organizations also usually have a better understanding of their performance (Porter and Tanner, 1998) and consequently take the necessary actions to improve it. Both the evidence and our experience suggest that managers are happy with the use of QMSs, such as Baldrige, European Foundation for Quality Management (EFQM), and ISO standards, and they are willing to continue with these approaches. Due to the deployment of a QMS, customers, employers, shareholders, and society benefit in several ways.

In a practical way, there is a great challenge for organizations to effectively translate a QMS into improvement actions and real benefits. In many cases, although organizations seek to build a good QMS, their efforts are locally deployed, and not systematically coordinated. The results may be good, but insufficient to stand out of the crowd and to fully benefit from the investments. Consequently, we strongly recommend that organizations seek professional consultancy before embarking on any QMS framework, to understand what their real needs and requirements are. This will increase the chances of implementing the right quality methods and tools and avoiding the potential pitfalls.

Coming back to the question: Does the QMS ensure business success? We can state that with a well-planned and executed QMS in your organization, you can achieve higher sales, increase profits, improve productivity, and enhance overall business performance. However, it is also necessary to have strong discipline and good decision making to select the right quality methods and tools and to deploy them efficiently and effectively. Many quality methods and fads have come and go, and some of them, in the words of experts, have been successful or unsuccessful. However, it can be argued that no single quality management initiative, model, or framework can guarantee any organization's success at any level. Ultimately, it is the ability of leaders and the commitment of top management to effectively translate any frameworks and strategies into real benefits.

9.5 Summary

This chapter briefly summarized what was reviewed about QMSs in this book. It then emphasized that a quality culture is not a requirement, strictly speaking, but a way of life for any organization that aims to produce and deliver high-standard quality products and services. Strong leadership and a shared vision are essential to accomplish this objective. Then the chapter discussed continual process improvement as a means of continuing to work toward business excellence. Finally, it stated the attributes of a successful organization and discussed the importance of a QMS to support that journey toward excellence and a high business performance that ultimately help achieve the strategic goal of becoming a world-class organization.

9.5.1 Key Points to Remember

- Make quality culture a way of doing business and a way of life in the organization.
- Clearly identify what a successful organization is and develop a strategic quality plan to get there.
- Promote the concept of a learning organization and implement the programs to ensure that its members understand the successes and failures.
- Plan and deploy continuous improvement programs.
- Have a vision, share it, and work hard to make things happen.

References

Marash, S. A., Berman, P., and Flying, M. (2004). *Fusion management.* QSU Publishing Company, Fairfax, VA.

Porter, L. J., and Tanner, S. J. (1998). *Assessing business excellence.* Butterworth-Heinemann, Woburn, MA.

Further Suggested Reading

Collins, J. (1994). *Built to last: Successful habits of visionary companies.* Harper Business, New York.

Collins, J. (2001). *Good to great.* Harper Business, New York.

Hamel, G., and Prahalad, C. K. (1996). *Competing for the future*. Harvard Business Review Press, Boston.

Peters, T. J., and Waterman, R. H. (2004). *In search of excellence: Lesson's from America's best run companies*, 2nd ed. Profile Books, London.

Index

Lightning Source UK Ltd.
Milton Keynes UK
UKHW032008171220
375320UK00004B/61